Introduction to
Differential Geometry
for Engineers

MONOGRAPHS AND TEXTBOOKS IN
PURE AND APPLIED MATHEMATICS

67. *J. K. Beem and P. E. Ehrlich*, Global Lorentzian Geometry (1981)
68. *D. L. Armacost*, The Structure of Locally Compact Abelian Groups (1981)
69. *J. W. Brewer and M. K. Smith, eds.*, Emmy Noether: A Tribute to Her Life and Work (1981)
70. *K. H. Kim*, Boolean Matrix Theory and Applications (1982)
71. *T. W. Wieting*, The Mathematical Theory of Chromatic Plane Ornaments (1982)
72. *D. B. Gauld*, Differential Topology: An Introduction (1982)
73. *R. L. Faber*, Foundations of Euclidean and Non-Euclidean Geometry (1983)
74. *M. Carmeli*, Statistical Theory and Random Matrices (1983)
75. *J. H. Carruth, J. A. Hildebrant, and R. J. Koch*, The Theory of Topological Semigroups (1983)
76. *R. L. Faber*, Differential Geometry and Relativity Theory: An Introduction (1983)
77. *S. Barnett*, Polynomials and Linear Control Systems (1983)
78. *G. Karpilovsky*, Commutative Group Algebras (1983)
79. *F. Van Oystaeyen and A. Verschoren*, Relative Invariants of Rings: The Commutative Theory (1983)
80. *I. Vaisman*, A First Course in Differential Geometry (1984)
81. *G. W. Swan*, Applications of Optimal Control Theory in Biomedicine (1984)
82. *T. Petrie and J. D. Randall*, Transformation Groups on Manifolds (1984)
83. *K. Goebel and S. Reich*, Uniform Convexity, Hyperbolic Geometry, and Nonexpansive Mappings (1984)
84. *T. Albu and C. Năstăsescu*, Relative Finiteness in Module Theory (1984)
85. *K. Hrbacek and T. Jech*, Introduction to Set Theory, Second Edition, Revised and Expanded (1984)
86. *F. Van Oystaeyen and A. Verschoren*, Relative Invariants of Rings: The Noncommutative Theory (1984)
87. *B. R. McDonald*, Linear Algebra Over Commutative Rings (1984)
88. *M. Namba*, Geometry of Projective Algebraic Curves (1984)
89. *G. F. Webb*, Theory of Nonlinear Age-Dependent Population Dynamics (1985)
90. *M. R. Bremner, R. V. Moody, and J. Patera*, Tables of Dominant Weight Multiplicities for Representations of Simple Lie Algebras (1985)
91. *A. E. Fekete*, Real Linear Algebra (1985)
92. *S. B. Chae*, Holomorphy and Calculus in Normed Spaces (1985)
93. *A. J. Jerri*, Introduction to Integral Equations with Applications (1985)
94. *G. Karpilovsky*, Projective Representations of Finite Groups (1985)
95. *L. Narici and E. Beckenstein*, Topological Vector Spaces (1985)
96. *J. Weeks*, The Shape of Space: How to Visualize Surfaces and Three-Dimensional Manifolds (1985)
97. *P. R. Gribik and K. O. Kortanek*, Extremal Methods of Operations Research (1985)
98. *J.-A. Chao and W. A. Woyczynski, eds.*, Probability Theory and Harmonic Analysis (1986)
99. *G. D. Crown, M. H. Fenrick, and R. J. Valenza*, Abstract Algebra (1986)
100. *J. H. Carruth, J. A. Hildebrant, and R. J. Koch*, The Theory of Topological Semigroups, Volume 2 (1986)

101. *R. S. Doran and V. A. Belfi*, Characterizations of C*-Algebras: The Gelfand-Naimark Theorems (1986)
102. *M. W. Jeter*, Mathematical Programming: An Introduction to Optimization (1986)
103. *M. Altman*, A Unified Theory of Nonlinear Operator and Evolution Equations with Applications: A New Approach to Nonlinear Partial Differential Equations (1986)
104. *A. Verschoren*, Relative Invariants of Sheaves (1987)
105. *R. A. Usmani*, Applied Linear Algebra (1987)
106. *P. Blass and J. Lang*, Zariski Surfaces and Differential Equations in Characteristic p > 0 (1987)
107. *J. A. Reneke, R. E. Fennell, and R. B. Minton*. Structured Hereditary Systems (1987)
108. *H. Busemann and B. B. Phadke*, Spaces with Distinguished Geodesics (1987)
109. *R. Harte*, Invertibility and Singularity for Bounded Linear Operators (1988).
110. *G. S. Ladde, V. Lakshmikantham, and B. G. Zhang*, Oscillation Theory of Differential Equations with Deviating Arguments (1987)
111. *L. Dudkin, I. Rabinovich, and I. Vakhutinsky*, Iterative Aggregation Theory: Mathematical Methods of Coordinating Detailed and Aggregate Problems in Large Control Systems (1987)
112. *T. Okubo*, Differential Geometry (1987)
113. *D. L. Stancl and M. L. Stancl*, Real Analysis with Point-Set Topology (1987)
114. *T. C. Gard*, Introduction to Stochastic Differential Equations (1988)
115. *S. S. Abhyankar*, Enumerative Combinatorics of Young Tableaux (1988)
116. *H. Strade and R. Farnsteiner*, Modular Lie Algebras and Their Representations (1988)
117. *J. A. Huckaba*, Commutative Rings with Zero Divisors (1988)
118. *W. D. Wallis*, Combinatorial Designs (1988)
119. *W. Więsław*, Topological Fields (1988)
120. *G. Karpilovsky*, Field Theory: Classical Foundations and Multiplicative Groups (1988)
121. *S. Caenepeel and F. Van Oystaeyen*, Brauer Groups and the Cohomology of Graded Rings (1989)
122. *W. Kozlowski*, Modular Function Spaces (1988)
123. *E. Lowen-Colebunders*, Function Classes of Cauchy Continuous Maps (1989)
124. *M. Pavel*, Fundamentals of Pattern Recognition (1989)
125. *V. Lakshmikantham, S. Leela, and A. A. Martynyuk*, Stability Analysis of Nonlinear Systems (1989)
126. *R. Sivaramakrishnan*, The Classical Theory of Arithmetic Functions (1989)
127. *N. A. Watson*, Parabolic Equations on an Infinite Strip (1989)
128. *K. J. Hastings*, Introduction to the Mathematics of Operations Research (1989)
129. *B. Fine*, Algebraic Theory of the Bianchi Groups (1989)
130. *D. N. Dikranjan, I. R. Prodanov, and L. N. Stoyanov*, Topological Groups: Characters, Dualities, and Minimal Group Topologies (1989)

Other Volumes in Preparation

Introduction to Differential Geometry for Engineers

BRIAN F. DOOLIN
Lockheed Missiles and Space Company
Sunnyvale, California

CLYDE F. MARTIN
Texas Tech University
Lubbock, Texas

MARCEL DEKKER, INC.   New York • Basel • Hong Kong

Library of Congress Cataloging--in--Publication Data

Doolin, B. F.
  Introduction to differential geometry for engineers / Brian F.
Doolin, Clyde F. Martin.
    p.   cm. -- (Monographs and textbooks in pure and applied
mathematics: 136)
  Includes bibliographical references and index.
  ISBN 0-8247-8396-4
  1. Geometry, Differential. I. Martin, Clyde. II. Title.
III. Series.
QA641.D66   1990
516.3'6-- --dc20   90-41747
                                                          CIP

This book is printed on acid-free paper.

MARCEL DEKKER, INC.
270 Madison Avenue, New York, New York 10016

Current printing (last digit):
10 9 8 7 6 5 4 3 2 1

PRINTED IN THE UNITED STATES OF AMERICA

To our eight children:
   the four Doolins—Anne, Clare, David and Beth
   the four Martins—Terri, Robert, David and Beth

Preface

This book has been written to acquaint engineers, especially control engineers, with the basic concepts and terminology of modern global differential geometry. The ideas discussed are applied here mainly as an introduction to the Lie theory of differential equations and to the role of Grassmannians in control systems analysis. To reach these topics, the fundamental notions of manifolds, tangent spaces, vector fields, and Lie algebras are discussed and exemplified. An appendix reviews such concepts needed for vector calculus as open and closed sets, compactness, continuity, and derivative.

Although the content is mathematical, this is not a mathematical treatise. Several excellent introductions to modern differential geometry exist, but they are written for readers with a strong mathematical, rather than engineering, background. Reading this book should help an engineer to read those treatises, as well as to understand the points, if not the detailed arguments, of research papers on geometric control, and many of those on nonlinear control.

We are indebted to many people for the content of this book: our colleagues at NASA, especially George Meyer, where these notes originated; Robert Hermann who convinced us of the utility of differential geometry; and to Roger Brockett who saw how to relate modern mathematics and control and was able to convince many of us that in order to do good control theory one needed to be able to do good mathematics.

We are also indebted to our families and friends who bore with us during the fourteen years from lectures to book. We would like to thank Elizabeth Martin who did an outstanding job of preparing the final manuscript.

Brian F. Doolin

Clyde F. Martin

Contents

List of Figures

Chapter 1

INTRODUCTION

This book presents some basic concepts, facts of global differential geometry, and some of its uses to a control engineer. It is not a mathematical treatise; the subject matter is well developed in many excellent books, for example, in references [1], [2], and [3], which, however, are intended for the reader with an extensive mathematical background. Here, only some basic ideas and a minimum of theorems and proofs are presented. Indeed, a proof occurs only if its presence strongly aids understanding. Even among basic ideas of the subject, many directions and results have been neglected. Only those needed for viewing control systems from the standpoint of vector fields are discussed.

Differential geometry treats of curves and surfaces, the functions that define them, and transformations between the coordinates that can be used to specify them. It also treats the differential relations that stitch pieces of curves or surfaces together or that tell one where to go next.

In thinking of functions that can define surfaces in space,

one is likely to think of real functions (functions assigning a real number to a given point of their argument) of three-space variables such as the kinetic energy of a particle, or the distribution of temperature in a room. Differential geometry examines properties inherent in the surfaces these function define that, of course, are due to the sources of energy or temperature in the surroundings. Or, given enough of these functions, one might use them as proper coordinates of a problem. Then the generalities of differential geometry show how to operate with them when they are used, for example, to describe a dynamic evolution.

Differential geometry, in sum, derives general properties from the study of functions and mappings so that methods of characterization or operation can be carried over from one situation to another. Global differential geometry refers to the description of properties and operations that are over 'large' portions of space.

Though the studies of differential geometry began in geodesy and in dynamics where intuition can be a faithful guide, the spaces now in this geometry's concern are far more general. Instead of considering a set of three or six real functions on a space of vectors of three or six dimensions, spaces can be described by longer ordered strings of numbers, by sets of numbers ordered in various ways, or by ordered sets of products of numbers. Examples are $n-$dimensional vector spaces, matrices, or multilinear objects like tensors. It is not just these sets of numbers, but also the rules one has of passing from one set to another that form the proper subject matter of differential geometry, linking it to matters of interest in control.

All analytic considerations of geometry begin with a

space filled with stacks of numbers. Before one can proceed to discuss the relations that associate one point with another or dictate what point follows another, one has to establish certain ground rules. The ground rules that say if one point can be distinguished from another, or that there is a point close enough to wherever you want to go, are referred to as topological considerations. The basic description of the topological spaces underlying all the geometry of this paper is given in an appendix on fundamentals of vector calculus. This appendix discusses such desired topological characteristics as compactness and continuity, which is needed to preserve these characteristics in passing from one space to another. The appendix concludes by recalling two theorems from vector calculus that provide the basic glue by which manifolds, the word for the fundamental spaces of global differential geometry, are assembled. Since this discussion is fundamental to differential geometry, we briefly review it. The review is relegated to an appendix, however, because it is not the topic of this book, nor should one dwell on it.

The first two chapters of the body of the book describe manifolds, the spaces of our geometry. Some simple manifolds are mentioned. Several definitions are given, starting with one closest to intuition then passing to one perhaps more abstract, but actually less demanding to verify in cases of interest in control engineering. Then mappings between manifolds are considered. A special space, the tangent space, is discussed in chapter 3. A tangent space is attached to every point in the manifold. Since this is where the calculus is done, it and its relations to neighboring tangent spaces and to the manifold that supports it must be carefully described.

Computation in these spaces is the topic of the next few chapters. Calculus on manifolds is given in chapter 4 on vector fields and their algebra, where the connection between global differential geometry and linear and nonlinear control begins to become clear. Chapters 5 and 6, which treat some algebraic rules, conclude our exposition of the fundamentals of the geometry.

The examples given as the development unfolds should not only help the reader understand the topic under discussion but should also provide a basic set for testing ideas presented in the current literature. Chapter 7 is intended to give the reader a glimpse of the structure supporting the spaces in which linear control operates. More comprehensive applications of differential geometry to control are given in the final major chapter of the paper.

Chapter 2

MANIFOLDS AND THEIR MAPS

The first part of this chapter is devoted to the concept of a manifold. It is defined first by a projection then by a more useful though less intuitive definition. Finally, it is seen how implicitly defined functions give manifolds. Examples are considered both to enhance intuition and to bring out conceptual details. The idea of a manifold is brought out more clearly by considering mappings between manifolds. The properties of these mappings occupy the last part of this chapter.

2.1 Differentiable Manifolds

Although the detailed global description of a manifold can be quite complicated, basically a differentiable manifold is just a topological space (X, Ω) that in the neighborhood of each point looks like an open subset of $\mathbf{R}^k$. (In the notation (X, Ω), X is some set and Ω consists of all the sets defined as open in X and that characterize its topology. As to the notation $\mathbf{R}^k$, each point in $\mathbf{R}^k$ is specified as an ordered set of k real numbers. These and other notions arising below are discussed in the appendix.) This description can be formalized into a definition:

Definition 2.1 *A subset M of $\mathbf{R}^n$ is a $k-$dimensional manifold if for each $x \in M$ there are: open subsets U and V of $\mathbf{R}^n$ with $x \in U$, and a diffeomorphism f from U to V such that:*

$$f(U \cap M) = \{y \in V : y^{k+1} = \cdots = y^n = 0\}$$

Thus, a point y in the image of f has a representation like:

$$y = (y^1(x), y^2(x), \cdots, y^k(x), 0, \cdots 0)$$

A straight line is a simple example of a one-dimensional manifold, a manifold in $\mathbf{R}^1$. It is a manifold in $\mathbf{R}^1$ even if it is given, for example, in $\mathbf{R}^2$. There it might represent the surface of solutions of the equation of a particle of unit mass under no forces: $\ddot{x} = 0$ and with given initial momentum: $\dot{x}(t = 0) = a$. In the coordinate system $y_1 = x; y_2 = \dot{x} - a$, the manifold is given by the points $(y_1, 0)$. To the particle, its whole world looks like part of $\mathbf{R}^1$ though we see its tracks clearly as part of $\mathbf{R}^2$. Any open subset of the straight line is also a one-dimensional manifold, but a closed subset of it is not.

The sphere in $\mathbf{R}^3$ is an example of a two-dimensional manifold. It is an example of a closed manifold and is often denoted as S^2. Thus, for a point P in $\mathbf{R}^3$: $P = (x_1, x_2, x_3)$, the manifold is given as the set:

$$S^2 = \{P \in \mathbf{R}^3 : x_1^2 + x_2^2 + x_3^2 - 1 = 0\}$$

Its two-dimensional character is clear when a point in S^2 is given in terms of two variables, say, latitude and longitude. Another map of S^2 into $\mathbf{R}^2$ is given by stereographic projections. Since this map not only has historical interest but also will be used later, it will now be discussed to show that S^2 is a two-dimensional manifold.

Let $U(r; P)$ be an r-neighborhood of a given point P of $\mathbf{R}^3$ in $M = S^2$ such that $U \cap M$ is the set

$$U \cap M = \{P : x_1^2 + x_2^2 + x_3^2 - 1 = 0; x_1^2 + x_2^2 \geq e(1 + x_3); e > 0\}$$

Then define the stereographic projection of a point P into the plane as the function from $U \cap M$ to $\mathbf{R}^3$:

$$f_{e,P} = (u_1, u_2, 0) = \left(\frac{x_1}{1 - x_3}, \frac{x_2}{1 - x_3}, 0 \right)$$

The mapping is illustrated in sketch (a), where the following ratios can be seen to hold: $u_2/x_2 = d/\ell$ and $\ell/d = (1 - x_3)/1$. The projection is generated by drawing a line from the 'North Pole' $(0,0,1)$ to a point on the sphere and continuing the line to the plane $x_3 = 0$. Thus, a point of the sphere is associated with a point on the plane and vice versa.

The map is written $f_{e,P}$ to call attention to the important role that the parameter e plays in restricting its domain of definition. With the restriction, f can be shown to be a diffeomorphism; without it, the function is not.

The one function is not enough to map the whole manifold. The point $(0, 0, 1)$ and some e neighborhood of it

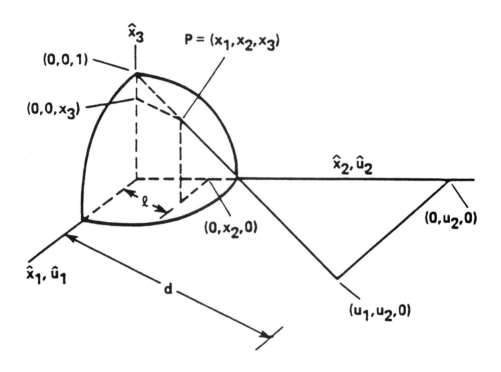

Figure 2.1: Sketch (a)

on the manifold have been excluded. Another similar map that includes these points but excludes others can be given by a stereographic projection from the 'South Pole'

$$g_{e,P} = \left(\frac{x_1}{1 + x_3}, \frac{x_2}{1 + x_3}, 0 \right)$$

with $g_{e,P}$ defined on the set

$$\{P : x_1^2 + x_2^2 + x_3^2 - 1 = 0; x_1^2 + x_2^2 \geq e(1 - x_3); e > 0\}$$

If the parameter e is given the value unity, f maps the lower hemisphere and g maps the upper hemisphere onto the interior of the unit circle on the plane that coincides with the plane $x_3 = 0$. Agreeably, the points of the sphere, where $x_3 = 0$, go into the same points of u_1 and u_2 under both maps.

The examples of one- and two-dimensional manifolds so far have been sets given in some $\mathbf{R}^n$ and mapped into $\mathbf{R}^1$ or $\mathbf{R}^2$. Sets forming manifolds are not always described naturally in some $\mathbf{R}^n$. To embed them in an $\mathbf{R}^n$ before showing that the definition is satisfied may be an undesirably awkward task. In fact, it is not necessary, and we will extend our previous definition so as to avoid it. That labor, however, will be avoided only at the expense of our introducing more formalism now.

Let M be a second countable, Hausdorff topological space. A *chart* in M is a pair (V, α) with V an open set and α a C^∞ function onto an open set in $\mathbf{R}^n$ and having a C^∞ inverse. A C^∞ *atlas* is a set of such charts, $\{(V_i, \alpha_i)\} = A$, with the following properties:

(i) $M = \cup V_i$

(ii) If (V_1, α_1) and (V_2, α_2) are in A and $V_1 \cap V_2 \neq \phi$, then

$$\alpha_2 \alpha_1^{-1} : \alpha_1(V_1 \cap V_2) \to \alpha_2(V_1 \cap V_2)$$

is a C^∞ diffeomorphism.

Sketch (b), which illustrates the subsets V_1 and V_2 and maps α_1 and α_2 may aid in picturing the content of condition (ii). With this formalism established, our second definition of a manifold can now be given:

Definition 2.2 *A C^∞ manifold is a pair (M, A) where M is second countable Hausdorff topological space, and A is a maximal C^∞ atlas.*

The conditions on the topology guarantee that the number of charts required to cover M is countable. The word 'maximal' gives a technical condition. It makes the atlas the class of collections of just enough charts to form a countable basis of charts. By referring to the class, one is not tied to a representation given by a particular set of charts.

Although the definition seems unduly complicated, it turns out to be just what is necessary to meet our intuition. Every m-dimensional manifold determined by the definition can in fact be considered as a subset of $\mathbf{R}^n$ for some $n : m \leq n \leq 2m + 1$. Any weakening of the definition can allow objects which cannot be embedded in some $\mathbf{R}^n$.

We opened this discussion of differentiable manifolds with the remark that basically a differentiable manifold is a topological space that in the neighborhood of each point looks like an open subset of $\mathbf{R}^k$. The first definition said that

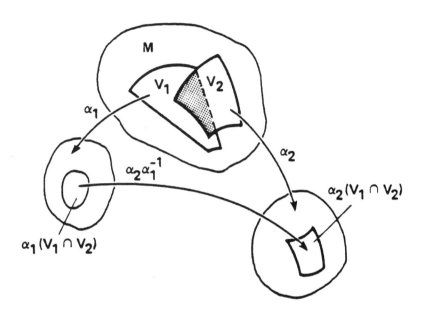

Figure 2.2: Sketch (b)

each neighborhood, even though expressed as a subset of $\mathbf{R}^n$, was equivalent to $\mathbf{R}^k$. That is, the space expressed in $\mathbf{R}^n$ really only had k, not n, degrees of freedom. Another way of saying this is by saying that a k-dimensional manifold can be expressed using n variables with $n - k$ conditions imposed on them.

These remarks are made because, in practice, manifolds are often given as the set of points where a certain function vanishes. The implicit function theorem gives conditions under which the vanishing of the function gives k constraints (exchanging the k and $n - k$ of the previous paragraph), so that only $n - k$ of the variables are free, and the space is a manifold with dimension $n - k$ if the theorem is satisfied everywhere. Then the manifold is said to be given implicitly, or by the implicit function theorem.

Formalizing the above remarks, we consider a C^∞ function F with domain $A \subseteq \mathbf{R}^n$ and range in $\mathbf{R}^k$. That is, for every choice of n real numbers $(x_1, \cdots, x_n)$ in A, the function F has the k real numbers $F = (f_1, \cdots, f_k)$. Let M be the set

$$M = \{x : F(x) = 0 = (0, 0, \cdots, 0)\}$$

If the rank of the Jacobian matrix F' is equal to k for all $x \in M$, then M is an $n - k$-dimensional manifold.

Under the conditions stated, the implicit function theorem says that k of the variables can be expressed in terms of the other $n - k$, and the latter can be given values arbitrarily. Another statement of the implicit function theorem (see [4, p. 43]) shows that a coordinate transformation can be found that assigns the value zero to the k explicit functions. In other words, the conditions of the first definition of a manifold are satisfied.

2.2 Examples

Consider the real function $F = a_1 x_1 + a_2 x_2 + a_3 x_3 - b = 0$. It is clear that $a_1 x_1 + a_2 x_2 + a_3 x_3 - b = 0$ describes a plane, a two-dimensional manifold, in $\mathbf{R}^3$. It is not difficult to imagine a change of coordinates that reorients $\mathbf{R}^3$ so that every point in the given plane can be written as $(y_1, y_2, 0)$, satisfying the first definition for a two-dimensional manifold. One also sees that the Jacobian matrix of F is $F' = (a_1, a_2, a_3)$, which has rank one for all x in $F(x) = 0$. The implicit function theorem, then, says the manifold is of dimension $3 - 1 = 2$.

Another example of using the implicit function theorem is given by the two-dimensional manifold S^2. Here $F = x_1^2 + x_2^2 + x_3^2 - 1 = 0$ and $F' = (2x_1, 2x_2, 2x_3)$. Now, F' is not zero because not all x_i vanish simultaneously, for $F = 0$ is not satisfied by $x_1 = x_2 = x_3 = 0$, Thus, F' has rank one and the manifold S^2 has dimension $3 - 1 = 2$.

Let a second condition be imposed on the space. For instance, consider the circle resulting from passing a plane containing the origin through S^2. In particular, consider the function

$$F = (f_1, f_2) : \mathbf{R}^3 \to \mathbf{R}^2$$

with f_1 as before,

$$f_1 = x_1^2 + x_2^2 + x_3^2 - 1 = 0$$

and

$$f_2 = x_1 - ax_2 = 0$$

The manifold $M = \{x : F = 0\}$ is the circle S^1. The

Jacobian matrix of F is:

$$F' = \begin{pmatrix} 2x_1 & 2x_2 & 2x_3 \\ 1 & -a & 0 \end{pmatrix}$$

the rank of which is two everywhere on the manifold. The dimension of this manifold, therefore, equals 3 - 2, or 1.

On the other hand, consider the function $G : \mathbf{R}^3 \to \mathbf{R}^1$ defined by:

$$G(x_1, x_2, x_3) = x_1^2 + x_2^2 - x_3^2$$

The zero set of G is a cone, but note that the rank of G' is 0 at $(0,0,0)$. Thus, at this point the cone doesn't satisfy the condition for the implicit function theorem, our definition of a manifold, or our intuitive view of its being locally like $\mathbf{R}^2$.

Exercise 1 *Consider the circle given by the equation $(x - 2)^2 + z^2 = 1$. If this circle is rotated about the $z-axis$ the resulting locus is a torus, usually denoted by $\mathbf{T}^2$. Show that a function defining the torus is given by*

$$f(x, y, z) = (x^2 + y^2 + z^2)^2 - 16(x^2 + y^2).$$

Show that $\{(x, y, z) : f(x, y, z) = (x^2 + y^2 + z^2 - 5)^2 - 16(1 - z^2)\} = \mathbf{T}^2$ is a manifold defined by the implicit function theorem.

To exercise our second definition of a manifold, let us consider S^2 again with charts defined as follows:

$$V_1 = \{(x_1, x_2, x_3) : x_3 \neq 1\}$$

$$V_2 = \{(x_1, x_2, x_3) : x_3 \neq -1\}$$

$$\alpha_1(x_1, x_2, x_3) = (u_1, u_2) = \left(\frac{x_1}{1 - x_3}, \frac{x_2}{1 - x_3}\right)$$

$$\alpha_2(x_1, x_2, x_3) = (v_1, v_2) = \left(\frac{x_1}{1 + x_3}, \frac{x_2}{1 + x_3}\right)$$

Then the set

$$A = \{(V_1, \alpha_1), (V_2, \alpha_2)\}$$

satisfies the first condition for an atlas. Consideration of the geometry of the stereographic projection shows that α_1 and α_2, with their domains restricted appropriately actually map onto $\mathbf{R}^2$ and are one to one. Also

$$\alpha_1(V_1 \cap V_2) = R^2 - \{(0,0)\} = \alpha_2(V_1 \cap V_2)$$

where $\mathbf{R}^2 - \{(0,0)\}$ means that the point $(0,0)$ has been deleted from $\mathbf{R}^2$. Furthermore, after calculating the inverse:

$$\alpha_1^{-1}(u_1, u_2) = \left(\frac{2u_1}{u_1^2 + u_2^2 + 1}, \frac{2u_2}{u_1^2 + u_2^2 + 1}, \frac{u_1^2 + u_2^2 - 1}{u_1^2 + u_2^2 + 1}\right)$$

one can see that:

$$\alpha_2\alpha_1^{-1}(u_1, u_2) = (v_1, v_2) = \left(\frac{u_1}{u_1^2 + u_2^2}, \frac{u_2}{u_1^2 + u_2^2}\right)$$

Since its partial derivatives exist and are continuous whenever $(u_1, u_2) \neq (0,0)$, $\alpha_2\alpha_1^{-1}$ is a C^∞ diffeomorphism. That a similar calculation yields the same conclusion for $\alpha_1\alpha_2^{-1}$ confirms that A is an atlas.

Thus, S^2 satisfies our new definition of a manifold. In fact, a little thought makes one realize that any space that is a manifold by the first definition is also one by the second. Furthermore, anything given as a manifold by the implicit function theorem satisfies both definitions.

Exercise 2 *Let* $S = \{(x,y) : -1 \le x < 1, \ -1 \le y < 1 \}$.
Define the pairs (S_1, ϕ_1), (S_2, ϕ_2), (S_3, ϕ_3), *and* (S_4, ϕ_4),
$\phi_i : S_i \to \mathbf{R}$, *by the following.*

$$S_1 = \{(x,y) : -1 < x < 1, \ -1 < y < 1 \}$$
$$\phi_1(x,y) = (x,y)$$

$$S_2 = \{(x,y) : -1 < x < 1, \ -1 \le y < 0, \ 0 < y < 1 \}$$
$$\phi_2(x,y) = \begin{cases} (x,y) & y < 0 \\ (x, y-2) & y > 0 \end{cases}$$

$$S_3 = \{(x,y) : \ -1 \le x < 0, \ 0 < x < 1, \ -1 < y < 1 \}$$
$$\phi_3(x,y) = \begin{cases} (x,y) & x < 0 \\ (x-2, y) & x > 0 \end{cases}$$

$$S_4 = \{(x,y) : \ -1 \le x < 0, \ 0 < x < 1, \ -1 \le y < 0, \ 0 < y < 1 \}$$
$$\phi_4(x,y) = \begin{cases} (x,y) & -1 \le x < 0, \quad -1 \le y < 0 \\ (x-2, y-2) & 0 < x < 1, \quad 0 < y < 1 \\ (x-2, y) & -1 \le y < 0, \quad 0 < x < 1 \\ (x, y-2) & -1 \le x < 0, \quad 0 < y < 1 \end{cases}$$

Show that the set $\{(S_1, \phi_1), (S_2, \phi_2), (S_3, \phi_3), (S_4, \phi_4)\}$ *is an atlas for the set* S *and hence that* S *has a* C^∞ *manifold structure. The point of this exercise is that a manifold may be given in more than one way. Note that what* (S_2, ϕ_2) *does is to identify the upper boundary with the lower boundary.*

Intuitively we are giving **S** *the structure of a torus.*

Another trivial but important example of a class of manifolds is afforded by any open subset of $\mathbf{R}^n$. There the atlas may consist of the set itself, together with the identity map. Thus, the notion that manifolds are spaces that locally look like open subsets of $\mathbf{R}^n$ is at least self-consistent. This example is important because the whole idea of the definition of manifolds is to be able to see how calculations valid in $\mathbf{R}^n$ carry over into any other manifold.

Another example of a manifold, which is an open set of Euclidean space and which is important in systems theory, follows. Let

$$\dot{x} = Ax + bu$$

be a single-input controllable system. Recall that controllability is equivalent to having the rank of the matrix

$$[b, Ab, A^2 b, \cdots, A^{n-1} b]$$

equal to n where A is an $n \times n$ matrix. Now let M be the set of pairs (A, b) such that a system is controllable:

$$M = \{(A, b) : \dot{x} = Ax + bu \text{ is controllable}\}$$

The complement of this set is the set that satisfies the condition:

$$det[b, Ab, A^2 b, \cdots, A^{n-1} b] = 0$$

Since this is a closed set in $\mathbf{R}^{n^2+n}$, the set M is open in $\mathbf{R}^{n^2+n}$ and therefore is a manifold.

The system, being of single input, is a special case. In general, when the control distribution function B is an $n \times m$ matrix, M^* is also a manifold where M^* is the set:

$$M^* = \{(A, B) : \dot{x} = Ax + Bu \text{ is controllable}\}$$

Although the conditions are more involved and less easy to describe than the determinant condition above, a similar argument shows that the controllable pairs are an open subset of $\mathbf{R}^{n(n+m)}$.

A more general example along these same lines is the set of triples of matrices (A, B, C) representing the system

$$\dot{x} = Ax + Bu$$

$$y = Cx$$

If the system is controllable and observable, it can be shown that this set of triples is also an open subset of a suitable Euclidean space.

Related to this manifold is a set of matrix transfer functions $T(s)$. These are matrices of rational functions that arise as the Laplace transforms of the above systems. If this set $\{T(s)\}$ is a manifold is a deep question in systems theory. It has been answered affirmatively by Martin Clark, [5], and Roger Brockett, [6], and independently by Michiel Hazewinkel, [7], and by Christopher Byrnes and N. Hurt, [8]. Much of the study in linear systems is involved with various properties of this manifold.

2.3   Manifold Maps

We have described manifolds and seen a few examples of them. Now we can describe the requirements on functions that allow them to be maps between manifolds, say, the manifolds M and N.

A function f,

$$f : M \to N$$

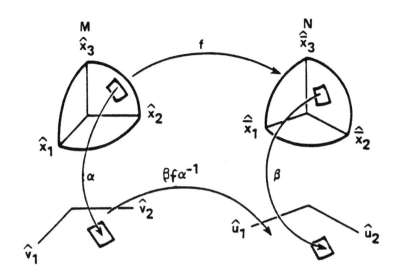

Figure 2.3: Sketch (c)

is a manifold map if for every $x \in M$ and chart (V, α) with $x \in V$, there is a chart (U, β) for N with $f(V) \subset U$ such that the composite function $\beta \circ f \circ \alpha^{-1}$:

$$\beta \circ f \circ \alpha^{-1} : \alpha(V) \to \beta(U)$$

is a C^∞ diffeomorphism. The relations are illustrated in sketch(c) for $f = A_1$, the map in the following example.

As an example, let $M = N = S^2$. Let A be a matrix such that $I = A^T A$, an orthogonal transformation. Then, if $x \in S^2$,

$$\mid Ax \mid^2 = (Ax) \cdot (Ax) = x^T A^T A x = x^T x = 1$$

Thus, A maps S^2 onto S^2. Consider a specific A, namely A_1:

$$A_1 = \begin{bmatrix} \sqrt{1/2} & \sqrt{1/2} & 0 \\ -\sqrt{1/2} & \sqrt{1/2} & 0 \\ 0 & 0 & 1 \end{bmatrix}$$

and let α and β be the previous stereographic projection α_1. Those relations become for α:

$$\alpha(x_1, x_2, x_3) = (v_1, v_2) = \left(\frac{x_1}{1 - x_3}, \frac{x_2}{1 - x_3} \right)$$

and

$$x_1 = v_1(1 - x_3) = \frac{2v_1}{v_1^2 + v_2^2 + 1}$$

$$x_2 = \frac{2v_2}{v_1^2 + v_2^2 + 1}$$

$$x_3 = \frac{v_1^2 + v_2^2 - 1}{v_1^2 + v_2^2 + 1}$$

The map A_1 gives $\bar{x} = A_1 x$:

$$\bar{x}_1 = \frac{1}{\sqrt{2}}(x_1 + x_2)$$

$$\bar{x}_2 = \frac{1}{\sqrt{2}}(-x_1 + x_2)$$

$$\bar{x}_3 = x_3$$

and $\beta A \alpha^{-1}$ maps $(v_1, v_2) \rightarrow (u_1, u_2)$ thus:

$$u_1 = \frac{\bar{x}_1}{1 - \bar{x}_3} = \frac{1}{\sqrt{2}} \frac{x_1 + x_2}{1 - x_3} = \frac{1}{\sqrt{2}}(v_1 + v_2)$$

$$u_2 = \frac{\bar{x}_2}{1 - \bar{x}_3} = \frac{1}{\sqrt{2}} \frac{-x_1 + x_2}{1 - x_3} = \frac{1}{\sqrt{2}}(-v_1 + v_2)$$

Both α and β are C^∞ functions on their respective domains. The composite map $(v_1, v_2) \to (u_1, u_2)$ is also clearly C^∞. The map A_1, therefore is a manifold map.

Thus, orthogonal transformations are manifold maps. They also form a manifold. The set of $n \times n$ orthogonal matrices form an $(n/2)(n-1)$ dimensional manifold. The truth of this statement will be verified by using the implicit function theorem.

The defining relation of an orthogonal matrix can be used to give a function of the matrices into the zero set:

$$f(X) = X^T X - I = 0$$

It must now be shown that the rank of the derivative is constant over all elements of the set. The derivative at $X = A$ can be found through the definition:

$$\lim_{\|H\| \to 0} \frac{1}{\|H\|} [f(X + H) - f(X) - f'(X)(H)]_{X=A} = 0$$

where the differential $f'(A)(H)$ recognizes that the derivative evaluated at A is a linear operator on H. Performing the expansion and considering the limit gives for the differential:

$$f'(A)(H) = H^T A + A^T H$$

Now A is invertible by definition and therefore, in the $n \times n$ case, maps one-to-one onto $\mathbf{R}^{n^2}$. Any matrix H is thus the image of some matrix under A, and one can write $H \to AH$. The derivative then gives:

$$f'(A)(AH) = H^T + H$$

Thus, the range of the derivative is seen to be the set of symmetric $n \times n$ matrices, which implies that it has the constant rank $(n/2)(n+1)$. Following the considerations of the implicit function theorem, the orthogonal $n \times n$ matrices form an $(n/2)(n-1)$–dimensional manifold.

A final consideration for this section is that of forming manifolds from the cartesian products of manifolds. If we have a manifold M with atlas A, we can construct a new manifold:

$$M \times M = \{(X,Y) : X, Y \in M\}$$

from M and A. The charts are constructed from the charts of A in the natural way as products, i.e., if (V_1, α_1) and (V_2, α_2) are charts in A. Then a chart for $M \times M$ is given by $(V_1 \times V_2, \alpha_1 \times \alpha_2)$ where

$$(\alpha_1 \times \alpha_2)(X,Y) = (\alpha_1(X), \alpha_2(Y))$$

Exercise 3 *Show that the torus $\mathbf{T}^2$ and the manifold $\mathbf{S}$ are diffeomorphic.*

Exercise 4 *Let $\mathbf{M} = \mathbf{S}^1 \times \mathbf{S}^1$. Show that $\mathbf{M}$ and $\mathbf{T}^2$ are diffeomorphic.*

Remark: In exercises 1,2,3 and 4, you have constructed the torus $\mathbf{T}^2$ in three quite different ways. Each construction has its merits and you can use any one of the constructions as the need arises. You will find that this often is the case in differential geometry and in general in mathematics. An object may have many concrete realizations, each of which has its particular uses. However, if you show something about one realization, it doesn't follow automatically that all realizations will share the shown property. Properties that all realizations share are sometimes referred to as *coordinate free* properties in geometry.

Chapter 3

TANGENT SPACES

The previous chapter defines manifolds and gives several examples of them. This chapter considers a basic construction of one manifold from another. While the method of construction itself is of interest insofar as it illustrates general procedures of modern differential geometry, the particular result, the tangent space, is an object of great importance. It is by way of the tangent space that calculus can be done in general situations.

To gain familiarity with the idea of a tangent space, it is worthwhile to spend some time with an example, that of the tangent space to the sphere. The information in the previous section concerning charts for the sphere allows charts to be constructed for this new space. The atlas resulting from the construction is examined in the light of the earlier definitions to see that this tangent space forms a manifold. The example is useful, too, for giving insight into such things as the dimensionality of a tangent space and the fact that its maps preserve its linear and differentiable structure. Part of the problem of constructing an atlas is that a map must be inverted and that its composition with another map be a diffeomorphism. Reducing our example from a sphere to a circle simplifies this calculation considerably.

Next, preparatory to considering the general construction of a tangent space, the notion of equivalence classes of curves on a manifold, and their addition and scalar multiplication is explored. This study provides the guide to the constructions that follow, and to the confirmation that the tangent space is a manifold.

The rest of the chapter is devoted to the tangent space in general. It is seen to be a manifold whose charts and chart maps are derived from those of the underlying manifold. It is seen to have vector space properties. Similar properties of maps between tangent manifolds are examined. The differentiating properties of these induced maps are noted.

3.1   The Tangent Space of Sphere

Consider an object moving on S^2, the surface of a sphere in space. As it moves, it generates a velocity vector which is tangent to the sphere at each point. Since it is tangent, the velocity lies in the tangent plane at each point and moves continuously through tangent planes as the object moves smoothly along the sphere. The concepts of a set of tangent planes and of smoothly transitioning from one to another is made precise by endowing the set of tangent planes together with their points of attachment with a manifold structure. That the set of tangent planes thus generates a manifold will be shown in two ways. The first uses the implicit function theorem. The second constructs charts and chart maps explicitly and shows that the chart maps have the requisite properties. This explicit construction reveals the relation of tangent spaces to derivative operations on the manifolds from which they are obtained.

Let the sphere, S^2, be described by the equation

$$x_1^2 + x_2^2 + x_3^2 = 1$$

and let the charts and chart maps be as given in the previous chapter. The tangent plane to S^2 at $\bar{x} = (x_1, x_2, x_3)$ is the set

$$T_x = \{(x_1 + y_1, x_2 + y_2, x_3 + y_3) : x_1 y_1 + x_2 y_2 + x_3 y_3 = 0\}$$

This is the space of vectors orthogonal to the radius vector (x_1, x_2, x_3) and translated to the origin. The manifold structure is obvious since there are two equations, one for the sphere, and the other for the tangent plane. That is, the tangent space, $T(S^2)$, is just the following set of points of $\mathbf{R}^6$:

$$\{(\bar{x}, \bar{y}) : x_1^2 + x_2^2 + x_3^2 - 1 = 0; x_1 y_1 + x_2 y_2 + x_3 y_3 = 0\}$$

The Jacobian matrix, therefore is:

$$\begin{bmatrix} 2x_1 & 2x_2 & 2x_3 & 0 & 0 & 0 \\ y_1 & y_2 & y_3 & x_1 & x_2 & x_3 \end{bmatrix}$$

Since not all the x_i vanish simultaneously, the rank of the matrix is 2 everywhere. The dimension of the manifold, therefore, is 4, corresponding to 2 degrees of freedom on the sphere, and 2 additional degrees of freedom on the plane tangent to the sphere at any point.

Now, if the object were moving freely in space ($\mathbf{R}^3$) and not constrained to the surface of the sphere, it would have three degrees of freedom of position. Its velocity vector, being unconstrained, would also have three degrees of freedom. As a manifold, then the tangent space would be made up of two copies of $\mathbf{R}^3$. That is to say, the tangent space of $\mathbf{R}^3$, $T(\mathbf{R}^3)$, is the cartesian product space $\mathbf{R}^3 \times \mathbf{R}^3$. If the object is considered to move in some open subset U of $\mathbf{R}^3$, then the space needed to describe all its possible positions and velocities is $U \times \mathbf{R}^3$. In this case, $T(U) = U \times \mathbf{R}^3$.

Return to considering motion on the sphere. From the chart maps of the previous section, we know that S^2 is locally like $\mathbf{R}^2$. It would be reasonable to suppose, then, that the tangent space of S^2 should look locally like $\mathbf{R}^2 \times \mathbf{R}^2$ as is confirmed by the calculations investigated in the following paragraphs. The calculations will consider a curve, c, on the sphere. To learn what is to be meant by the tangent, or velocity, of the curve at some point on the sphere, we will calculate the velocity of the image of the curve in $\mathbf{R}^2$, where we know what a tangent to a curve is. The image of c in $\mathbf{R}^2$ is obtained through the chart map, of course. Inspecting the form of the calculation of the tangent to the curve in $\mathbf{R}^2$ shows that it is the image of the tangent of the curve in the manifold under a mapping given by the derivative of the chart map. This inspection leads to the definition of the tangent to the curve on the manifold, and to the isolation of the map that maps it into $\mathbf{R}^2$.

Let c be a curve on the sphere; that is, $c(t)$ is the position of a point at t, where t is an element of an open interval in $\mathbf{R}^1$, and $t = 0$ is likewise in this interval. If it isn't, define a translation of $\mathbf{R}^1$ so that the zero does occur there. Furthermore, define the translation so that $c(0) = \bar{x}_0$. Let

(U, ϕ) be a chart containing $c(0)$. The chart map transforms the curve $c(t)$ by the composition $\phi \circ c$ into a curve in $\mathbf{R}^2$. The velocity of $\phi \circ c$ is just the derivative with respect to time: $(d/dt)\phi \circ c$. It is reasonable to expect that a chart map of the tangent space, $T(S^2)$, map the velocity vector of c into the velocity vector of $\phi \circ c$. Now, the derivative of $\phi \circ c$ at $t = 0$ is just:

$$\frac{d}{dt}(\phi \circ c)(0) = (\phi \circ c)'(0) = \phi'(\bar{x}_0)c'(0)$$

since the derivatives in question are defined. Thus, the velocity vectors c' are mapped by the derivative ϕ' of the chart map to elements of the 'local' tangent space $(\phi \circ c)'$. Take, for example, the chart map ϕ:

$$\phi(x_1, x_2, x_3) = \left(\frac{x_1}{1 - x_3}, \frac{x_2}{1 - x_3}\right)$$

then ϕ' is the matrix:

$$\phi'(x_1, x_2, x_3) = \begin{bmatrix} \frac{1}{1-x_3} & 0 & \frac{x_1}{(1-x_3)^2} \\ 0 & \frac{1}{1-x_3} & \frac{x_2}{(1-x_3)^2} \end{bmatrix}$$

A reasonable candidate for a chart in the tangent space $T(S^2)$ is then $[T(U), T\phi(x, y)]$;

$$T(U) = \{(x_1, x_2, x_3, y_1, y_2, y_3) : \bar{x}^2 - 1 = 0; \bar{x} \cdot \bar{y} = 0; x_3 \neq 1\}$$

$$\begin{aligned} T\phi(\bar{x}, \bar{y}) &= [\phi(\bar{x}), \phi'(\bar{x})\bar{y}] \\ &= \left[\frac{x_1}{1 - x_3}, \frac{x_2}{1 - x_3}, \frac{y_1}{(1 - x_3)^2} + \frac{x_1 y_3}{(1 - x_3)^2}, \right. \\ &\qquad \left. \frac{y_2}{1 - x_3} + \frac{x_2 y_3}{(1 - x_3)^2}\right] \end{aligned}$$

It can be seen that $T(S^2)$ belongs to $\mathbf{R}^2 \times \mathbf{R}^2$, as expected.

Similarly, another tentative chart for $T(S^2)$ can be derived from the other chart for S^2:

$$T(V) = \{(x_1, x_2, x_3, y_1, y_2, y_3) : \bar{x}^2 - 1 = 0; \bar{x} \cdot \bar{y} = 0;$$
$$x_3 \neq -1\}$$

$$T\gamma(\bar{x}, \bar{y}) = [\gamma(\bar{x}), \gamma'(\bar{x})\bar{y}]$$
$$= \left[\frac{x_1}{1 + x_3}, \frac{x_2}{1 + x_3}, \frac{y_1}{1 + x_3} - \frac{x_1 y_3}{(1 + x_3)^2}, \right.$$
$$\left. \frac{y_2}{1 + x_3} - \frac{x_2 y_3}{(1 + x_3)^2} \right]$$

For proof that these charts and maps form an atlas, it must be shown that the union of the sets $T(U)$ and $T(V)$ covers the tangent space, and that compositions of one map with the inverse of the other over common domains of definition lead to C^∞ maps. It is clear that the first condition is satisfied, $T(S^2) = T(U) \cup T(V)$. The remainder of the argument is tedious and is carried out only for the circle, for which it will be seen that $[T(U), T\phi]$ and $[T(V), T\gamma]$ actually form an atlas.

Although computing the composition map of, for example,

$$T\phi \circ (T\gamma)^{-1}(\bar{a}, \bar{b})$$

is tedious, it is important to note that the outcome has the form

$$T\phi \circ (T\gamma)^{-1}(\bar{a}, \bar{b}) = [\phi \circ \gamma^{-1}(\bar{a}, \bar{b}), L(\bar{a})\bar{b}]$$

where $L(\bar{a})$ is an invertible linear transformation for each $\bar{a}$ in the domain. Thus, the composite map preserves both the differentiable structure of the tangent space and its linear structure.

The computations are particularly simple for the circle, S^1, which will be obtained by restricting S^2 to the plane $x_1 = 0$. S^1 will be mapped onto $\mathbf{R}^1$ by ϕ and γ where the image of $\bar{x} = (x_2, x_3)$ will be u_1 and v_1, respectively. The image of the tangent in $S^1, \bar{y} = (y_2, y_3)$, will be u_2 and v_2, respectively. S^1 is the set:

$$\{(\bar{x}, \bar{y}) : x_2^2 + x_3^2 - 1 = 0; x_2 y_2 + x_3 y_3 = 0\}$$

Under the chart maps ϕ, γ :

$$u_1 = \phi(\bar{x}) = \frac{x_2}{1 - x_3}; \quad v_1 = \gamma(\bar{x}) = \frac{x_2}{1 + x_3}$$

The Jacobian matrices at $(\bar{x}, \bar{y})$ are:

$$\phi'(\bar{x}) = \left(\frac{1}{1 - x_3}, \frac{x_2}{(1 - x_3)^2} \right); \quad \gamma'(\bar{x}) = \left(\frac{1}{1 + x_3}, \frac{-x_2}{(1 + x_3)^2} \right)$$

giving the tangent vectors in $\mathbf{R}^1$:

$$u_2 = \phi'(x)\bar{y} = \frac{y_2}{1 - x_3} + \frac{x_2 y_3}{(1 - x_3)^2};$$

$$v_2 = \gamma'(\bar{x})\bar{y} = \frac{y_2}{1 + x_3} - \frac{x_2 y_3}{(1 + x_3)^2}$$

To compute $\phi^{-1}(u_1)$ and $\gamma^{-1}(v_1)$ requires use of $x_2^2 + x_3^2 - 1 = 0$. One finds:

$$x_2 = \frac{2u_1}{u_1^2 + 1} = \frac{2v_1}{v_1^2 + 1}; \quad x_3 = \frac{u_1^2 - 1}{u_1^2 + 1} = \frac{1 - v_1^2}{v_1^2 + 1}$$

for points in the domain common to ϕ and γ.

To compute $(\phi')^{-1}(u_1, u_2)$ and $(\gamma')^{-1}(v_1, v_2)$ requires use of both $x_2^2 + x_3^2 - 1 = 0$ and $x_2 y_2 + x_3 y_3 = 0$. One finds:

$$y_2 = \frac{-2u_2(u_1^2 - 1)}{(u_1^2 + 1)^2} = \frac{2v_2(1 - v_1^2)}{(v_1^2 + 1)^2};$$

$$y_3 = \frac{4u_1 u_2}{(u_1^2 + 1)^2} = \frac{-4v_1 v_2}{(v_1^2 + 1)^2}$$

Finally, the composition $T\phi \circ (T\gamma)^{-1}(\bar{v}) = [\phi \circ \gamma^{-1}(\bar{v}), \phi' \circ (\gamma')^{-1}(\bar{v})]$ is found to be:

$$u_1 = \frac{1}{v_1}; \ u_2 = \frac{-v_2}{v_1^2} = \frac{-1}{v_1^2} v_2$$

These expressions give a diffeomorphism, since the point $u_1 = v_1 = 0$ is excluded as not being in the images of $U \cap V$. Note that u_2 is given by a linear transformation at $\bar{v}$.

Exercise 5 *Calculate the equations of the tangent planes to the torus given by*

$$0 = f(x, y, z) = (x^2 + y^2 + z^2)^2 - 16(x^2 + y^2).$$

Define the tangent space of the torus, $\mathbf{T}(\mathbf{T}^2)$ explicitly.

Exercise 6 *Define the tangent space to the manifold $\mathbf{S}$ explicitly.*

3.2   Equivalence Classes of Curves

The example of a tangent space started with a sphere and a plane tangent to the sphere at a point. This plane was seen to contain the tangent to the velocity of a curve passing through the point of attachment. As a matter of fact, the plane is the locus of tangents to all the curves on the surface at the point, a curve being a map of an interval of $\mathbf{R}^1$ into some region of the manifold. The interval of the real line is so adjusted for discussion that $t = 0$ corresponds to the point p of the manifold: $c(0) = p$. The curves can be grouped into classes. Being a tangent vector is taken to be a class property, and the tangent plane can be determined by a set of independent tangent vectors.

The classes are equivalence classes. Two curves are in the same class if they are equivalent to each other. They are equivalent if they pass through the same point and have the same velocity there. The velocity is measured in the local Euclidean frame given by the chart map attached at the point. Thus, the curves $c_1(t)$ and $c_2(t)$ are equivalent if

$$\frac{d}{dt}(\phi \circ c_1) \mid_{\phi(p)} = \frac{d}{dt}(\phi \circ c_2)\mid_{\phi(p)}$$

This equality is also written as

$$(\phi \circ c_1)'(\phi(p)) = (\phi \circ c_2)'(\phi(p))$$

The set of all curves equivalent to c at p is denoted by $[c]_p$. This symbol of the class of curves includes the point of attachment as well as the tangent vector.

Note that a particular chart was used in the definition of equivalence. It must be shown that the definition is independent of the particular chart used. This independence is something that must be routinely verified in almost every definition of differential geometry. In this case, as often, the verification requires just a routine manipulation of derivatives. Suppose (V, ψ) is another chart at $p = c_1(0) = c_2(0)$, and that c_1 is equivalent to c_2:

$$(\phi \circ c_1)'(\phi(p)) = (\phi \circ c_2)'(\phi(p))$$

Then apply the chain rule of differentiation symbolically:

$$\begin{aligned}
(\psi \circ c_1)'(\phi(p)) &= (\psi \circ \phi^{-1} \circ \phi \circ c_1)'(\phi(p)) \\
&= ((\psi \circ \phi^{-1})'(\phi \circ c_1(0)) \circ (\phi \circ c_1)'(\phi(p)) \\
&= ((\psi \circ \phi^{-1})'(\phi \circ c_2(0)) \circ (\phi \circ c_2)'(\phi(p)) \\
&= (\psi \circ \phi^{-1} \circ \phi \circ c_2)'(\phi(p)) \\
&= (\psi \circ c_2)'(\phi(p))
\end{aligned}$$

Thus, the definition goes through with any chart map.

The set of tangent vectors at a point, $\{[c]_p\}$, can be seen to form a vector space once it is understood just how addition of curves and their multiplication by a constant works. The operations act on the derivatives: $[c_2]_p = a[c_1]_p$ means $(\phi \circ c_2)'(\phi(p)) = a(\phi \circ c_1)'(\phi(p))$, according to the definition. The common point, $p = c_1(0) = c_2(0)$, remains fixed.

Consider two curves $c_1(t)$ and $c_2(t)$ on a manifold, M, with I_1 and $I_2 \subset \mathbf{R}^1$ such that $t \in I_1 \to c_1(t), t \in I_2 \to c_2(t), t = 0 \in I_1 \cap I_2 \neq \phi$ (the zero of time occurs in the common interval which is not empty). Suppose $c_1 \in [c_1]_p$ and $c_2 \in [c_2]_p$, and the question is how to add them; that is how to define $(c_1 + c_2)(t)$.

The meaning of addition and scalar multiplication of curves is clear when the operations are defined in the local cartesian space. The definitions come out most easily when the chart maps map the point p of the manifold into the origin of the local $\mathbf{R}^n : \phi \circ c(0) = \phi(p) = 0$. Under these conditions, $\phi \circ c_1$ and $\phi \circ c_2$ are curves at 0 in $\mathbf{R}^n$. Hence,

$$(\phi \circ c_1 + \phi \circ c_2) : I_1 \cap I_2 \to \mathbf{R}^n$$

is also a curve there, and $\phi^{-1}(\phi \circ c_1 + \phi \circ c_2)$ is a curve at p. Then addition is defined by defining $[c_1]_p + [c_2]_p$ to be the equivalence class $[\phi^{-1}(\phi \circ c_1 + \phi \circ c_2)]_p$:

$$[c_1]_p + [c_2]_p \equiv [\phi^{-1}(\phi \circ c_1 + \phi \circ c_2)]_p = [c_1 + c_2]_p$$

That the identification of the sum of equivalence classes of curves as the equivalence class of the sum of curves is well defined, that is, is independent of the particular curves chosen from the class, is easily shown. Let $c_1, b_1 \in [c_1]_p$, and $c_2, b_2 \in [c_2]_p$.

$$\begin{aligned}
\frac{d}{dt}(\phi \circ c_1 + \phi \circ c_2)(0) &= \frac{d}{dt}\phi \circ c_1(0) + \frac{d}{dt}\phi \circ c_2(0) \\
&= \frac{d}{dt}\phi \circ b_1(0) + \frac{d}{dt}\phi \circ b_2(0) \\
&= \frac{d}{dt}(\phi \circ b_1 + \phi \circ b_2)(0)
\end{aligned}$$

Then

$$[\phi^{-1}(\phi \circ c_1 + \phi \circ c_2)]_p = [\phi^{-1}(\phi \circ b_1 + \phi \circ b_2)]_p$$

shows that addition is well defined. Multiplication by a scalar is also well defined.

3.3 The Tangent Space in General

The set of all equivalence classes of curves passing through the point p of a manifold is said to be its tangent space there:

$$T_p(M) \equiv \{[c]_p, \text{ for all } c(t) \in M \text{ with } c(0) = p\}$$

The collection of tangent spaces of all the points of the manifold is called the tangent space of the manifold: $T(M) \equiv \cup_p T_p(M)$. This tangent space is itself a manifold with a structure that maintains both the linear structure of the equivalence classes of curves and the differentiable structure of the manifold from which it comes.

During the discussion of the tangent space to the sphere, it was mentioned that any open set of it had the structure of the cartesian product $U \times \mathbf{R}^2$, where U was an open subset of the sphere, with the local appearance of $\mathbf{R}^2 \times \mathbf{R}^2$. This product structure of the tangent manifold is general and is understood to follow from its definition as a union of tangent spaces when the nature of the point $[c]_p$ in general is recalled. Since $[c]_p$ is the class of curves at p, it can be written (in the appealing form of a Taylor expansion, but with some equivocation of addition): $[p + c'(0)t]_p$. It is specified by the vectors $c(0) = p$ and $c'(0)$, and can be written as $[p, c'(0)]_p$. Thus, the identification of $\cup_p T_p(U)$ with $U \times \mathbf{R}^n$ follows from identifying $[c]_p$ with $[p, c'(0)]_p$.

Again, locally, one chart map for $T(S^2)$ was given as

$$T\phi(\bar{x}, \bar{y}) = [\phi(\bar{x}), \phi'(\bar{x})\bar{y}] \subset \mathbf{R}^2 \times \mathbf{R}^2$$

which in the present notation is $[0, \phi'c'(0)]$. To show that this form holds generally, let (U, ϕ) be a chart in M. Being an open subset of M, U is a manifold. Thus, $T(U)$ is a well defined subspace of $T(M)$. The corresponding map, $T\phi$, maps $T(U)$ into $T(\phi(U))$. That is to say,

$$T\phi[c]_p = [\phi \circ c]_{\phi(p)}$$

Taking $c(t) \in [c]_p$ in the form $c(t) = c(0) + c'(0)t$, one can expand ϕ similarly:

$$\phi(c(t)) = \phi(p + c'(0)t) = \phi(p) + \phi'c'(0)t$$

for small enough t. Since $(\phi \circ c)'(0)$ defines the equivalence class, one has

$$T\phi[c]_p = [\phi \circ c]_{\phi(p)}$$

as claimed, and:

$$T\phi : T(U) \to T(\phi(U)) = \phi(U) \times \mathbf{R}^n$$

Thus, the intuition obtained from the discussion of the sphere holds.

for $T\phi$ to be a map, it must be invertible, or, equivalently, 'one-to-one and onto'. The identification of $T(\phi(U))$ with $\phi(U) \times \mathbf{R}^n$ shows that it is onto. To verify that it is one to one, suppose that

$$T\phi[c]_p = T\phi[b]_q$$

which was just seen to mean

$$[\phi(p), (\phi \circ c)'(0)]_{\phi(p)} = [\phi(q), (\phi \circ b)'(0)]_{\phi(q)}$$

This implies that $\phi(p) = \phi(q)$, and since ϕ is one to one, $p = q$. This fact, together with $(\phi \circ c)' = (\phi \circ b)'$ and the definition of equivalence, show that $[c]_p = [b]_q$. Hence, $T\phi$ is one to one and onto an open set.

To investigate the compatibility and differentiability requirements of chart maps, let (V, γ) be another chart and map in M. construct

$$T\gamma : T(V) \to \gamma(V) \times \mathbf{R}^n$$

and assume $U \cap V \neq \{0\}$. the map $T\phi \circ (T\gamma)^{-1}$ is found as follows. Let $(a, b) \in \gamma(V) \times \mathbf{R}^n$. Then

$$
\begin{aligned}
T\phi \circ (T\gamma)^{-1}(a,b) &= \phi \circ [\gamma^{-1} \circ (a + bt)]_{\gamma^{-1}(a)} \\
&= \phi(\gamma^{-1}(a)), \frac{d}{dt}\phi \circ \gamma^{-1} \circ (a + bt)|_{t=0} \\
&= [\phi(\gamma^{-1}(a)),(\phi \circ \gamma^{-1})'(a)b]_{\phi(\gamma^{-1}(a))}
\end{aligned}
$$

Since we are assuming $\phi \circ \gamma^{-1}$ is C^∞, $T\phi \circ (T\gamma)^{-1}$ is also a C^∞ function. Thus, we have an atlas for $T(M)$ whose charts are derived from those of M and are given by $(T(U)), T\phi)$ and the like. Furthermore, note that the composite map is differentiable, and since $(\phi \circ \gamma^{-1})'$ is a linear map, it also respects the linear structure of the tangent space.

3.4   Tangent Space Maps

Our discussion of mapping between tangent spaces need not examine the requirements put on the maps themselves. They are the same as those between manifolds in general, and are found in the previous section. They are very similar to the requirements of chart maps, to which, of course, they must reduce when the map between the manifolds is the identity.

This same resemblance of the behavior of manifolds under chart maps between the global and local manifolds and the behavior of the tangent spaces under chart maps, is seen in the two properties to be looked at now. One of the points to be made is that the tangent manifold map contains the manifold map and its derivative in the same way that the tangent manifold chart map is seen to contain the manifold chart map and its derivative. The other point to be made is that the vector space structure of the tangent manifold is preserved under the maps. This property is due to the linearity of the derivative map.

The map between tangent spaces contains the map between manifolds and its derivative. These results can be obtained very quickly in just the same way that the same result is seen to hold for the tangent manifold chart maps. For simplicity, consider M and N to be open subsets of Euclidean space and $f : M \to N$ a manifold map. It was noted before that the equivalence classes of curves in M are represented by linear functions; that is,

$$[c]_p = [p + c'(0)t]_p$$

Now if $f : M \to N$ then

$$Tf[p + c'(0)t]_p = [f(p + c'(0)t)]_{f(p)} = [f(p) + tf'(p)c'(0)]_{f(p)}$$

The map of a particular curve, then, can be written as:

$$Tf(p, c'(0)) = (f(p), Df(p)c'(0)) \qquad (3.1)$$

For fixed p, then, the linear map is just $Df(p)$, and Tf contains both f and the derivative.

Consider for example, a rotation of the sphere. Let $M = N = S^2$. If f is a rotation of S^2, it can be represented by an orthonormal matrix, say, $f(x) = Ax$, and $AA^T = I$. Let c be a curve on S^2 with $c(0) = p = (x_1, x_2, x_3)$. If $c'(0) = (y_1, y_2, y_3)$, then c determines the point in the tangent space $c = (x_1, x_2, x_3, y_1, y_2, y_3)$. Note that the definition of the tangent space $T(S^2)$ requires that $p^T c'(0) = x_1 y_1 + x_2 y_2 + x_3 y_3 = 0$, which also has to hold in the image space after the rotation.

The rotation f determines a new curve $f \circ c$ at $f(p)$, and

$$(f \circ c)'(0) = (Ac)'(0)$$

The tangent space map, then looks like

$$Tf(p, c'(0)) = (Ap, Ac'(0))$$

where $A = Df(p)$ since a rotation is a linear map. To verify that the image of the map is also $T(S^2)$, note that

$$(Ap)^T Ac'(0) = p^T A^T Ac'(0) = p^T c'(0) = 0$$

as expected.

To examine the addition property under mappings of the tangent manifolds, again consider the manifold map $f : M \to N$. Let (U, ϕ) be a chart in M and (V, γ) be a chart in N. Assume that $f(U) \subset V$ (otherwise, take U' to be $f^{-1}(V) \cap U$). We will examine the action of Tf on a single tangent plane $T_p(M)$.

Let $p \in U$. We have an addition defined on $T_p(M)$ with respect to the chart (U, ϕ) and an addition defined on $T_{f(p)}(N)$ with respect to the chart (V, γ). We want to compare $Tf([c_1]_p + [c_2]_p)$ with $Tf[c_1]_p + Tf[c_2]_p$ with the hope that they are equal. The argument is typical: we reduce each expression, using definitions, until something rather obvious appears to connect terms:

$$\begin{aligned} Tf([c_1]_p + [c_2]_p) &= TF[\phi^{-1} \circ (\phi \circ c_1 + \phi \circ c_2)]_p \\ &= [f \circ \phi^{-1} \circ (\phi \circ c_1 + \phi \circ c_2)]_{f(p)} \end{aligned}$$

and

$$\begin{aligned} Tf[c_1]_p + Tf[c_2]_p &= [f \circ c_1]_{f(p)} + [f \circ c_2]_{f(p)} \\ &= [\gamma^{-1} \circ (\gamma \circ f \circ c_1 + \gamma \circ f \circ c_2)]_{f(p)} \end{aligned}$$

By definition, the final two equivalence classes are equal if and only if

$$\frac{d}{dt}\gamma \circ f \circ \phi^{-1} \circ (\phi \circ c_1(t) + \phi \circ c_2(t))|_{t=0}$$

$$= \frac{d}{dt}(\gamma \circ f \circ c_1(t) + \gamma \circ f \circ c_2(t))|_{t=0}$$

Now, calculating the derivative on the left yields

$$\frac{d}{dt}\gamma \circ f \circ \phi^{-1} \circ (\phi \circ c_1(t) + \phi \circ c_2(t))|_{t=0}$$

$$= (\gamma \circ f \circ \phi^{-1})'(0)[(\phi \circ c_1)'(0) + (\phi \circ c_2)'(0)]$$

$$= (\gamma \circ f \circ \phi^{-1})'(0)(\phi \circ c_1)'(0) +$$
$$(\gamma \circ f \circ \phi^{-1})'(0)(\phi \circ c_2)'(0)$$

$$= (\gamma \circ f \circ \phi^{-1} \circ \phi \circ c_1)'(0) +$$
$$(\gamma \circ f \circ \phi^{-1} \circ \phi \circ c_2)'(0)$$

$$= \frac{d}{dt}[\gamma \circ f \circ c_1(t) + \gamma \circ f \circ c_2(t)]|_{t=0}$$

Thus, we have the important fact that the addition goes through so that the restriction of Tf to the tangent planes is a linear function.

We might remark parenthetically that tangent spaces are a special case of the more general concept of vector bundles. A vector bundle is a triple of objects $(\pi, E, B,)$ where E and B are manifolds: π is a manifold mapping of E onto B, and $\pi^{-1}(b)$ is a vector space for each $b \in B$. In the case of tangent spaces, B is the manifold, E is the tangent space $T(B)$, and π is the map defined by $\pi([c]_p) = p$. The vector space $T(B)$ is called the *fiber* of E over b. The map π^{-1} is called a cross section when it is one-to-one and onto.

All the words and worries of this and the previous section should not obscure what are basically simple concepts. The wealth of discussion and terminology aims at separating the many ideas growing close together as topics in geometry and analysis grow. A simple example and some diagrams may keep the reader aware of the whole topic as details are described.

In a typical elementary discussion, the derivative of a polynomial might be shown as $(d/dx)(x^2) = 2x$. A later discussion would say that differentiating the function $x \rightarrow x^2$ gives the function $x \rightarrow 2x$. This statement can be shown diagrammatically like sketch (d).

The advantage is that the function can be viewed as an object having a structure and operations of its own. Though this looks more complicated than necessary, it does help resolve details of what is happening [9].

Similarly, the relationships of a mapping between manifolds, of their tangent spaces, and of the mapping between the tangent spaces induced by that between the manifolds can be illustrated like sketch (e), which shows the relation between the manifold map, $f : M \rightarrow N$ and the map thereby induced between the tangent spaces, $Tf : T(M) \rightarrow T(N)$. Looking back over the discussion of charts, one can see that the same sort of diagram illustrates the same relations between a chart map to the local $\mathbf{R}^n$ and that it induced in the tangent space (sketch(f)).

The maps X and Y in sketches (e) and (f) relate the manifolds M, N, and $\mathbf{R}^n$ to their respective tangent spaces. The discussion of these relations is the task of the next section.

Figure 3.1: Sketch(d)

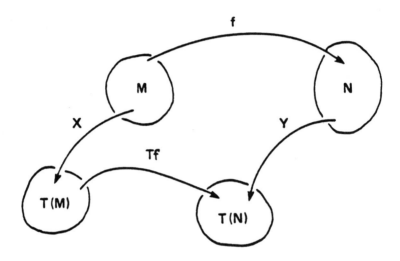

Figure 3.2: Sketch(e)

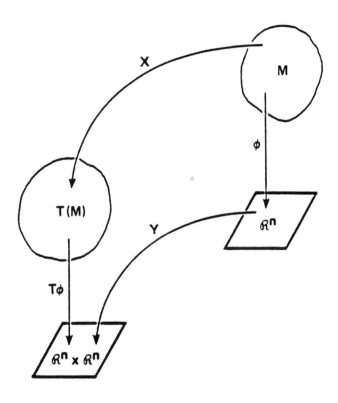

Figure 3.3: Sketch(f)

Exercise 7 *Show that the tangent space of T^2 is diffeomorphic to $T(S^1) \times T(S^1)$.*

Remark: You have shown that the tangent space construction for T^2 is independent of coordinates. Of course, this will generally be true.

Chapter 4

VECTOR FIELDS

So far, manifolds and their tangent spaces have been defined and discussed. It has been shown that the tangent spaces are related to velocities. Their fundamental relation to differential equations has been hinted at. This relation will become explicit in this chapter which examines vector fields as entities that relate tangent spaces to the manifolds underlying them.

Defining vector fields this way is a modern choice among alternatives. They actually have many familiar connections. Once one has started from one definition and from the fundamental properties coming directly from this definition, then the properties coming from the other connections have to be established. To help establish these properties, the notion of 'derivation'— the operation of taking derivatives — is brought in. This notion is first given abstractly, then interpreted in terms of the Euclidean plane. Some of its properties are easy to obtain. These are related to those of a vector field by showing that there is an isomorphism between spaces of vector fields and spaces of derivatives. finally, it is an important fact that vector fields not only form a space, but also an algebra. The last part of this chapter contains a discussion of a multiplication rule (symbolized by brackets) and some of the properties of the algebra coming from the multiplication, as well as an example using the linear state equations of control theory.

4.1   Vector Fields

Let M be a manifold and $T(M)$ its tangent space. A vector field, X, is a manifold map from M to $T(M)$ such that for every $p \in M$, the vector field at the point p gives a point in the tangent space attached to the manifold there: $X(p) \in T_p(M)$.

Recall that each point, t, on the smooth curve $c(t)$ has a velocity vector belonging to a class with tangent vector $c'(t)$, and that $T_p(M)$ is the set of equivalence classes of curves through the point p, at $t = 0$. A good choice of scale, then, gives $c'(t)$ as the point $(t, d/dt(t))$, or, $(t, 1)$. Now consider the curve c on M with domain I as a map between manifolds; namely, $c : I \to M$. Let $c'(t)$ denote the curve in $T_p(M) : c'(0) \in [c]_p$. As shown in sketch (g), there is also an induced map T_c between the tangent space to $I, I \times \mathbf{R}_1$ and $T(M) : Tc : I \times \mathbf{R}_1 \to T(M)$.

The commuting properties shown in sketch (g) indicate that $c'(t)$ is the image of the point $(t, 1)$:

$$c'(t) = Tc(t, 1) \tag{4.1}$$

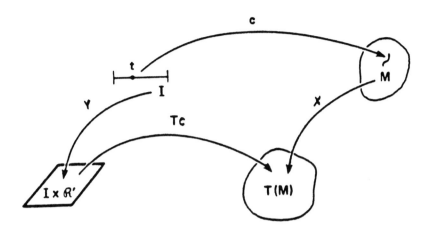

Figure 4.1: Sketch(g)

Now the definition of vector fields says that $c'(t)$ is the image of $c(t) \in M$ by the vector field X:

$$c'(t) = X(c(t)) \tag{4.2}$$

To look at this relation in more detail, consider it in local coordinate where M can be taken as an open subset of $\mathbf{R}^n$ and its tangent space, $T(M)$, a subset of $\mathbf{R}^n \times \mathbf{R}^n$. Equations (4.1) and (3.1) give the tangent vector $c'(t)$ as:

$$\begin{aligned} c'(t) &= Tc(t, 1) \\ &= (c(t), Dc(t) \cdot 1) \\ &= (c(t), \frac{d}{dt}c(t)) \end{aligned} \tag{4.3}$$

Now we can write $X(c(t))$ in the form:

$$X(c(t)) = (c(t),\ \underset{\sim}{X}\ (c(t))) \qquad (4.4)$$

A comparison of equations (4.2), (4.3), and (4.4) shows that;

$$\frac{d}{dt}c(t) =\ \underset{\sim}{X}\ (c(t)) \qquad (4.5)$$

This is a system of differential equations.

Equation (4.5) is important. It gives a conceptual identification of a moving point with a function of position. It shows that the thrust of the definition of vector fields is that the manifold is the locus over all initial conditions of all the curves whose tangent vectors are given by equation (4.5). They are the 'integral curves' of $\underset{\sim}{X}$.

The entire local theory of ordinary differential equations applies to the system of equations in (4.5). Its study in terms of manifolds gives rise to some global results. One example of a global result is given by the following theorem which will be used in a later section:

Theorem 4.1 *Let M be a compact manifold, X a vector field on M and $c : I \to M$ an integral curve of X. Then the domain of c can be extended to* **R**.

One interpretation of this theorem, for example, is that there is no finite escape time for solutions of differential equations on compact manifolds, which means that the solutions are well behaved over any finite time interval.

4.2 Derivations

Having defined vector fields and shown that they determine the right-hand side of ordinary differential equations,
we turn to the concept of derivations later to be shown
equivalent to that of a vector field. The concept is useful
also because it is usually easier to perform the derivations
than to calculate the velocity vector directly.

Let M be a manifold and let $F(M)$ be all the real-valued
functions that map M into $\mathbf{R}$. Since when any of them are
evaluated at a given point they are just real numbers, any
two functions $f, g \in F(M)$ can be added and multiplied
pointwise:

$$
\begin{aligned}
(f+g)(x) &= f(x) + g(x) \\
(fg)(x) &= f(x)g(x)
\end{aligned}
$$

Since the right-hand sides are familiar operations on real
numbers, they define the symbols on the left. (These definitions, together with statements about multiplying by constants and associative properties, make the set of real functions $F(M)$ into a 'ring'. An alternative approach to our
development of manifolds uses rings of functions defined on
open sets rather than charts because knowing $F(M)$ you
know M.)

A derivation, θ, on $F(M)$ is defined as a function that
maps $F(M)$ to $F(M)$, with the following properties:

1. $\theta(cf) = c\theta(f), c \in \mathbf{R}$

2. $\theta(f + g) = \theta(f) + \theta(g)$

3. $\theta(fg) = \theta(f)g + f\theta(g)$

Properties (1) and (2) say that θ is a linear map on $F(M)$. Property (3) shows that the operation is not linear when the coefficient of a function is not a constant but another function. It looks suspiciously like the usual derivative of the product of functions. It should come as no surprise, then, that these three properties defining the derivation operation, θ, abstract the usual idea of a derivative.

Now derivatives are differencing operations that need to have three objects specified before they give a value, say, a real number. They need a function to work on, a location at which the results can be evaluated, and a heading away from the point of evaluation. The statement $\theta f(x)$, or $\theta(f)(x)$, gives a real number. The function f and the point of evaluation, x, are explicit. Implicit in θ, then, are the notions of limits of differencing and of the direction of differencing.

Consider an example. Let M be the Euclidean plane: $M = \mathbf{R}^2$. Then take $F(\mathbf{R}^2)$ as the set of all real-valued functions with continuous partial derivatives of all orders. Let θ be given by:

$$\theta = \frac{\partial}{\partial x} + \frac{\partial}{\partial y}$$

Then

$$\theta(f) = \frac{\partial f}{\partial x} + \frac{\partial f}{\partial y} \tag{4.6}$$

and, by the ordinary rules of calculus, one has

$$\theta(fg) = \theta(f)g + f\theta(g)$$

A more general example of a derivation for $f \in F(\mathbf{R}^2)$ is

$$\alpha(f) = \alpha_1(x,y)\frac{\partial f}{\partial x} + \alpha_2(x,y)\frac{\partial f}{\partial y} \qquad (4.7)$$

where α_1 and α_2 also belong to $F(\mathbf{R}^2)$. The direction of differentiation is (α_1, α_2). In fact, every derivation of $F(\mathbf{R}^2)$ has the form of equation (4.7), as is seen from the following.

Consider a function $f \in F(\mathbf{R}^2)$. Then the equation

$$f(x,y) = f(a,b) + \int_0^1 \frac{\partial}{\partial t} f(a + t(x-a), b + t(y-b))dt$$

is an identity. It can be written as:

$$
\begin{aligned}
f(x,y) \;=\;& f(a,b) + \\
& (x-a)\int_0^1 \frac{\partial f}{\partial x}(a + t(x-a), b + t(y-b))dt \\
+\;& (y-b)\int_0^1 \frac{\partial f}{\partial y}(a + t(x-a), b + t(y-b))dt
\end{aligned}
$$

For a fixed a and b, this formula is true for all x and y. Letting

$$g_1(x,y) = \int_0^1 \frac{\partial f}{\partial x}(a + t(x-a), b + t(y-b))dt$$

and

$$g_2(x,y) = \int_0^1 \frac{\partial f}{\partial y}(a + t(x-a), b + t(y-b))dt$$

we can write

$$f(x,y) = f(a,b) + (x-a)g_1(x,y) + (y-b)g_2(x,y) \qquad (4.8)$$

Define two projection functions:

$$p_1(x, y) = x \text{ and } p_2(x, y) = y \qquad (4.9)$$

Then equation (9) can be written in function notation as

$$f = f(a, b) + (p_1 - a)g_1 + (p_2 - b)g_2$$

Now applying an arbitrary derivation, θ, to f gives:

$$
\begin{aligned}
\theta(f) &= \theta(p_1 g_1) - a\theta(g_1) + \theta(p_2 g_2) - b\theta(g_2) \\
&= \theta(p_1)g_1 + p_1\theta(g_1) - a\theta(g_1) + \theta(p_2)g_2 \quad (4.10) \\
&\quad + p_2\theta(g_2) - b\theta(g_2)
\end{aligned}
$$

since properties (1) and (3) give the condition that $\theta(c) = \theta(b) = 0$. Evaluating expression (4.10) at the point (a,b) gives

$$
\begin{aligned}
\theta(f)(a, b) &= \theta(p_1)(a, b)g_1(a, b) + a\theta(g_1)(a, b) \\
&\quad - a\theta(g_1)(a, b) \\
&\quad + \theta(p_2)(a, b)g_2(a, b) + \\
&\quad b\theta(g_2)(a, b) - b\theta(g_2)(a, b) \\
&= \theta(p_1)(a, b)g_1(a, b) + \theta(p_2)(a, b)g_2(a, b)
\end{aligned}
$$

Now

$$g_1(a, b) = \int_0^1 \frac{\partial f}{\partial x}(a, b)dt = \frac{\partial f}{\partial x}(a, b) \int_0^1 dt = \frac{\partial f}{\partial x}(a, b)$$

likewise

$$g_2(a, b) = \frac{\partial f}{\partial y}(a, b).$$

Thus, we have

$$\theta = \theta(p_1)\frac{\partial}{\partial x} + \theta(p_2)\frac{\partial}{\partial y}$$

which has the form of equation (4.7).

4.3 A Digression on Notation

The symbol θ has been introduced. It is about to be related to the symbol X, for a vector field, through the symbol L_X, for a Lie derivative. All relate to symbols used for elements of the tangent space. All will look like a common gradient or directional derivative when applied to real functions in ordinary Euclidean space. With this much notation being used to emphasize different aspects of basically the same set of objects, it seems desirable to digress from the development of ideas to a comparison of the forms under discussion.

Notational problems begin with the expression for the tangent spaces. A single tangent space, $T_p(M)$, is a linear vector space attached to a particular point p of a manifold. It has the structure of the product of the spaces: $M \times \mathbf{R}^n$; that is, a tangent vector can be written as (p, x), with $p \in M$ and $x \in \mathbf{R}^n$. All the vector operations are done with the second component, but they only make sense if they are done at the same point of space: $(p, x) + (p, y) = (p, x + y)$. Whenever vector operations of tangent vectors are discussed, it is assumed that the objects of the operation reside at the same point of the manifold, even if this requirement is not reverted to explicitly. They are not defined as otherwise.

The explicit expression for the $\mathbf{R}^n$ term has been written variously as $c'(0), (f \circ c(0))', f'(0)c'(0)$, and so on, depending on obvious circumstances. Strictly speaking, these expression refer to the velocity of the curve at a point in the manifold, which is equivalent to a tangent vector in the tangent space, a distinction not always kept clear.

The velocity aspects of this term, which refers to its source in a curve on a manifold, may not always be significant; the tangent living in $\mathbf{R}^n$ is. It might be referred to as Df when something generic is meant. It might take the form $df(m)(m, x)$ in mapping an element (m, x) from one tangent space to another. This form emphasizes the linear operator aspects of $df(m)$. If the original manifold is Euclidean, then $df(m)(m, x)$ can be written explicitly as $(\partial f / \partial x^i(m), x^i)$ where the outer parentheses and the comma denote an inner product. This can also look like $(\partial f / \partial x^i(m), \alpha^i)$, where its connection with the general form of $\theta f = \alpha^i(\partial f / \partial x^i)$ is clear. This is also the general form for a directional derivative. The same form represents $X(f)$ under these circumstances. It is also the form that a Lie derivative takes when it operates on a scalar function.

4.4 The Isomorphism

It has already been admitted that vector fields and derivations are just different sides of the same coin. Their identification with each other is formalized by an isomorphism between them. It will be shown that each vector field on a manifold gives rise to a unique derivation of $F(M)$, and with the result of every algebraic operation between vector fields there corresponds the result of an algebraic operation between the corresponding derivations.

Recall that for $f \in F(M)$, the range of the induced map Tf is $\mathbf{R} \times \mathbf{R}$:

$$Tf : T(M) \to T(R) = \mathbf{R} \times \mathbf{R}$$

Also recall that restricting its domain to the neighborhood of $T_m(M)$

$$T_m f = Tf | T_M(M)$$

was shown earlier to be a linear map:

$$T_m f : T_m(M) \to \{f(m)\} \times \mathbf{R}$$

Define $df(m)$ by:

$$df(m) = p_2 T_m f$$

where p_2 is the projection onto the second coordinate as defined in equation (10): $p_2(a, b) = b$.

When M is an open subset of $\mathbf{R}^n$ so that $T(M) = \mathbf{R}^n \times \mathbf{R}^n$, then f is a C^∞ real-valued function of n real variables. The induced map at the point (m, x) of the tangent space gives:

$$T_m f(m)(m, x) = (f(m), Df(m)x) \in \mathbf{R}^1 \times \mathbf{R}^1$$

from which the number $df(m)(m, x)$ can be identified:

$$df(m)(m, x) = p_2 T_m f(m)(m, x) = Df(m)x$$

It can be seen from these expressions that $df(m)$ results from the linear operator $Df(m)$ acting on X.

Since $Df(m)$ is a linear map from $\mathbf{R}^n \to \mathbf{R}^1$, where it gives a linear functional, it is an element of a space dual to $T_m(M)$, and can be represented as a row of n elements. This row notation is quite compatible with the explicit notation for a derivative usually to be found in texts of advanced calculus:

$$Df(m) = \frac{\partial f}{\partial x^1}(m), \cdots, \frac{\partial f}{\partial x^n}(m) \qquad (4.11)$$

If $x^i(x)$ is the i'th coordinate function for the Euclidean manifold,

$$df(m)(m, dx) = \frac{\partial f}{\partial x^1}(m)dx^1 + \cdots + \frac{\partial f}{\partial x^n}(m)dx^n = Df(m)dx$$
$$(4.12)$$

which formalizes the usual expression in advanced calculus texts:

$$df = \frac{\partial f}{\partial x^1}dx^1 + \cdots + \frac{\partial f}{\partial x^n}dx^n = (dx^1\frac{\partial}{\partial x^1} + \cdots + dx^n\frac{\partial}{\partial x^n})f$$

The expressions in (4.11) and (4.12) are worth a second glance in light of the comment that $Df(m)$ is an element of the space dual to the tangent space $T_m(M)$. Though the terms in the right-hand side of (4.12) are real numbers, their factors come from (4.11) and a column of dx^i. If the dx^i are considered to be unit vectors in the dual space, (4.12) is an element of that space. According to the usual method of evaluating the coefficients of a vector or of a dual (or co–) vector, the coefficients of (4.12) must come from the action of unit elements in the original vector space to which the dx^i are dual. It does no violence to any interpretation of the usual differential expression to take these unit vectors to be the $\partial/\partial x^i$.

The above view sees expressions like (4.12) in a new light. It allows the introduction of another notation, that of the Lie derivative of a real function, f, with respect to a vector field X. This is written:

$$L_X(f)(m) = df(m)(X(m)) \tag{4.13}$$

This Lie derivative form will prove to be a notational convenience whose use can be interchanged with the differential operator form. Since the two forms shift different elements in or out of parentheses and exchange the written order of the factors, the choice of form used will be dictated largely by the desire for brevity of notation, and, of course, by which of the factors are relevant at the time.

The definition in (4.13) is pointwise: it holds at each point in the tangent and dual, or cotangent spaces associated with the point m. Since L_X is going to give the isomorphism between X and θ, $L_X f$ will have to be an element of $F(M)$, and will have to be extended from the point definition in (4.13). To show $L_X f \in F(M)$, one proves that df is a smooth map. This follows after the cotangent space is proved to be a manifold. The details are lengthy and technical. The interested reader can refer to [3] for the construction of the cotangent space and for the proof that $L_X f \in F(M)$. This then establishes that L_X maps from $F(M)$ to $F(M)$.

To see that L_X is linear, recall what has been shown

about $T_m f$ and remember in particular that

$$T_m f : T_m(M) \to \mathbf{R} \times \mathbf{R}.$$

Equation (4.13) gives:

$$L_X(f + g)(m) = d(f + g)(m)(X(m))$$

Consider

$$
\begin{aligned}
d(f & + g)(m)[c]_m \\
&= p_2 \cdot T_m(f + g)[c]_m \\
&= p_2((f + g)(m), D((f + g) \circ c)(0)) \\
&= D(f \circ c)(0) + D(g \circ c)(0) \\
&= p_2(f(m), D(f \circ c)(0) + p_2(g(m), D(g \circ c)(0)) \\
&= p_2 \cdot T_m f[c]_m + p_2 \cdot T_m g[c]_m \\
&= (p_2 \cdot T_m f + p_2 \cdot T_m g)[c]_p \\
&= (df(m) + dg(m))[c]_m
\end{aligned}
$$

So d is additive and the rest of linearity is easy.

The same procedure that showed that L_X is a linear operator on $F(M)$ can be used to show that it is a derivation:

$$
\begin{aligned}
p_2 \cdot T_m fg[c]_m &= p_2(fg(m), D(fg \circ c)(0)) \\
&= D(fg \circ c)(0) \\
&= D(f \circ cg \circ c)(0) \\
&= f \circ c(0)D(g \circ c)(0) + g \circ c(0)D(f \circ c)(0) \\
&= f(m)D(g \circ c)(0) + g(m)D(f \circ c)(0) \\
&= f(m)p_2 \cdot T_m g[c]_m + g(m)p_2 \cdot T_m f[c]_m
\end{aligned}
$$

Thus, $L_X(fg) = fL_X(g) + gL_X(f)$, showing that each vector field, X, determines a derivation on $F(M)$.

Also note that from what we know of $T_m f$ the following holds:

$$L_{X+fY}(g) = L_X(g) + fL_Y(g)$$

so that, when acting on $F(M)$, L is a linear map from the set of all vector fields to the vector space of derivations.

A linear map from one vector space to another is an isomorphism if it is one-to-one and onto. The present case is one-to-one if $L_X f = 0$ for all f means that $X = 0$. To show that involves showing that every element in the dual space of $T_m(M)$ is represented by some $df(m)$. The existence proof is a 'partitions-of-unity' construction which can be found in [3]. Once the dual space is known to be represented by some $df(m)$, the proof that L is one-to-one simple. For, consider $L_X = 0$. Then $L_X(f)(m) = 0$ for all f in $F(M)$. Then

$$df(m)(X(m)) = 0$$

holds and every element α in the dual of $T_m(M)$ has the property that $\alpha(X(m)) = 0$. Thus, $X(m) = 0$, and this is true for each m in M.

The mapping is onto if every derivation arises from the Lie derivative of a vector field. Recall that for $\mathbf{R}^2$ we showed that every derivation θ has the form

$$\theta(f)(m) = \alpha_1(m)\frac{\partial f}{\partial x}|_m + \alpha_2(m)\frac{\partial f}{\partial y}|_m$$

Consider the vector field $X(m) = (m, \alpha_1(m), \alpha_2(m))$; then:

$$L_X(f)(m) \;=\; df(m)(X(m))$$
$$=\; \alpha_1(m)\frac{\partial f}{\partial x}|_m + \alpha_2(m)\frac{\partial f}{\partial y}|_m$$

and thus $L_X = \theta$. In the case, then, that $M = \mathbf{R}^2$, we have proved the theorem:

Theorem 4.2 *The linear space of vector fields on M is isomorphic to the linear space of derivations of $F(M)$.*

The general case is similar, essentially showing that on M the theorem can be proved locally using the proof given here.

4.5   The Algebra

The composition of two derivation is not generally a derivation for:

$$\theta_2\theta_1(fg) \;=\; \theta_2[f\theta_1 g + g\theta_1 f]$$
$$=\; \theta_2 f\theta_1 g + \theta_2 g\theta_1 f + f\theta_2\theta_1 g + \qquad (4.14)$$
$$g\theta_2\theta_1 f$$

The first two terms spoil the derivation property. On the other hand, note:

$$\theta_1\theta_2(fg) = \theta_1 f\theta_2 g + \theta_1 g\theta_2 f + f\theta_1\theta_2 g + g\theta_1\theta_2 f \qquad (4.15)$$

Subtracting equation (16) from equation (15) gives:

$$(\theta_2\theta_1 - \theta_1\theta_2)(fg) = f(\theta_2\theta_1 - \theta_1\theta_2)g + g(\theta_2\theta_1 - \theta_1\theta_2)f$$

so that the operation $\theta_2\theta_1 - \theta_1\theta_2$ is a derivation. This difference operation is called the commutator, or bracket, of θ_2 and θ_1, and is represented thus:

$$\theta_2\theta_1 - \theta_1\theta_2 = [\theta_2, \theta_1]$$

Using the Lie derivative form as an example to show that the bracket of Lie derivatives is a Lie derivative, consider a two-dimensional case with

$$L_X f \triangleq a^1(x)\frac{\partial}{\partial x^1}f + a^2(x)\frac{\partial}{\partial x^2}f = \left(\sum_j (a(x))^j \frac{\partial}{\partial x^j}\right)f$$

and

$$L_Y f \triangleq b^1(x)\frac{\partial}{\partial x^1}f + b^2(x)\frac{\partial}{\partial x^2}f = \left(\sum_j (b(x))^j \frac{\partial}{\partial x^j}\right)f$$

Then

$$(L_X L_Y)f \qquad\qquad (4.16)$$

$$= \left(a^1\frac{\partial}{\partial x^1} + a^2\frac{\partial}{\partial x^2}\right)\left(b^1\frac{\partial}{\partial x^1} + b^2\frac{\partial}{\partial x^2}\right)f$$

$$= \left(\left(a^1\frac{\partial b^1}{\partial x^1} + a^2\frac{\partial b^1}{\partial x^2}\right)\frac{\partial}{\partial x^1} + \left(a^1\frac{\partial b^2}{\partial x^1} + a^2\frac{\partial b^2}{\partial x^2}\right)\frac{\partial}{\partial x^2}\right.$$

$$+b^1\left(a^1\frac{\partial^2}{\partial x^1 \partial x^1} + a^2\frac{\partial^2}{\partial x^2 \partial x^1}\right)$$

$$+b^2\left(a^1\frac{\partial^2}{\partial x^1 \partial x^2} + a^2\frac{\partial^2}{\partial x^2 \partial x^2}\right)\Bigg)f \qquad (4.17)$$

and

$$(L_Y L_X)f \qquad (4.18)$$

$$= \left(b^1\frac{\partial}{\partial x^1} + b^2\frac{\partial}{\partial x^2}\right)\left(a^1\frac{\partial}{\partial x^1} + a^2\frac{\partial}{\partial x^2}\right)f$$

$$= \left(\left(b^1\frac{\partial a^1}{\partial x^1} + b^2\frac{\partial a^1}{\partial x^2}\right)\frac{\partial}{\partial x^1} + \left(b^1\frac{\partial a^2}{\partial x^1} + b^2\frac{\partial a^2}{\partial x^2}\right)\frac{\partial}{\partial x^2}\right.$$

$$+a^1\left(b^1\frac{\partial^2}{\partial x^1 \partial x^1} + b^2\frac{\partial^2}{\partial x^2 \partial x^1}\right)$$

$$+a^2\left(b^1\frac{\partial^2}{\partial x^1 \partial x^2} + b^2\frac{\partial^2}{\partial x^2 \partial x^2}\right)\Bigg)f \qquad (4.19)$$

Then

$$(L_X L_Y - L_Y L_X)f \qquad (4.20)$$

$$= [L_X, L_Y]f$$

$$= \left(\left(a^1\frac{\partial b^1}{\partial x^1} - b^1\frac{\partial a^1}{\partial x^1} + a^2\frac{\partial b^1}{\partial x^2} - b^2\frac{\partial a^1}{\partial x^2}\right)\frac{\partial}{\partial x^1}\right.$$

$$+ \left(a^1\frac{\partial b^2}{\partial x^1} - b^1\frac{\partial a^2}{\partial x^1} + a^2\frac{\partial b^2}{\partial x^2} - b^2\frac{\partial a^2}{\partial x^2}\right)\frac{\partial}{\partial x^2}\Bigg)f$$

$$= \left(\sum_j\left(\sum_i\left(a^i\frac{\partial b^j}{\partial x^i} - b^i\frac{\partial a^j}{\partial x^i}\right)\right)\frac{\partial}{\partial x^j}\right)f$$

$$= \left(\sum_j (c(x))^j\frac{\partial}{\partial x^j}\right)f = L_Z f \qquad (4.21)$$

Bracketing removes the terms with higher derivatives of f, leaving an expression with the proper form for a Lie derivative.

The isomorphism which the Lie derivative gives between vector fields and derivations means that operations involving derivations have corresponding operations involving vector fields. The operation on vector fields corresponding to the bracketing operation on derivations is also called a bracket and is written similarly. If X and Y are the vector fields corresponding to L_X and L_Y, then there is a vector field Z corresponding to $[L_x, L_Y]$ such that $Z = [X, Y]$. The following theorem holds:

Theorem 4.3 *The bracket operation on the linear space of vector fields forms a Lie algebra with the following properties:*

$$\left. \begin{array}{ll} (i) & [X, Y] = -[Y, X] \\ (ii) & [X, Y + aZ] = [X, Y] + a[X, Z], a \in \mathbf{R}^1 \\ (iii) & [X, [Y, Z]] + [Y, [Z, X]] + [Z, [X, Y]] = 0 \end{array} \right\} \quad (4.22)$$

A Lie algebra is, in fact, nothing more than a set of elements that forms a vector space and for which a multiplication is defined by a bracketing operation with the three properties in the theorem.

4.6 An Example of a Lie Algebra

The state space representation of linear constant-coefficient control systems is a good source of an example of a Lie algebra. From the control system

$$\dot{x} = Ax + Bu$$

we obtain a family of vector fields, one for each constant u. Let Z_u be the vector field that acts on a point x by the following:

$$Z_u(x) = (x, (Ax + Bu)) \in T_x(M) \subset \mathbf{R}^n \times \mathbf{R}^n.$$

Suppressing the base point x, let $Z_u = Ax + Bu$. The example of a Lie algebra will involve calculating the smallest Lie algebra that contains all the elements Z in $T_x(M)$. Define $Z_u = (X + U)$ with $U = Bu$ and $X = Ax = Z_0$. Also from equations (4.22),

$$
\begin{aligned}
[Z_u, Z_v] &= [X + U, X + V] \\
&= -[X, U] + [X, V] + [U, V]
\end{aligned}
$$

where the fact that $[X, X] = 0$ for any X has been used. Thus, the Lie algebra generated by the Z_u is given by the sets $\{X\}$, which is a singleton set, and $\{U : U \in \mathbf{R}^n\}$. These brackets can be evaluated by calculating the brackets of the corresponding Lie derivatives. The derivatives are written like that in equation (4.21):

$$L_X(f)(x) = \left(\sum_j (Ax)^j \frac{\partial}{\partial x^j} \right) f$$

and

$$L_U(f)(x) = \left(\sum_j (Bu)^j \frac{\partial}{\partial x^j} \right) f$$

Calculating the bracket for $[L_U, L_V]f = L_U L_V f - L_V L_U f$, one sees that all the second partials cancel, as shown in equations (4.17) and (4.19). Furthermore, since the coefficients $(Bu)^j$ do not involve x, their derivatives vanish. Hence, $[L_U, L_V] = 0$. The next bracket to consider is $[L_X, L_U]f$, the first terms of which is $L_X L_U$:

$$
\begin{aligned}
L_X L_U f &= L_X \sum_i (Bu)^i \frac{\partial}{\partial x^i} f \\
&= \sum_j \sum_i (Ax)^j \left(\frac{\partial}{\partial x^j} (Bu)^i \right) \frac{\partial}{\partial x^i} f + S \\
&= S
\end{aligned}
\tag{4.23}
$$

where S represents the second-order partial derivatives, the other terms vanishing. The other term of the bracket is a little more complex (take b_j^i as the elements of B and a_j^i as the elements of A):

$$
\begin{aligned}
L_U L_X f &= L_U \sum_i (Ax)^i \frac{\partial}{\partial x^i} f \\
&= \sum_j \sum_i (Bu)^j \frac{\partial}{\partial x^j} \left((Ax)^i \frac{\partial}{\partial x^i} f \right) \\
&= \sum_j \sum_i \sum_k \sum_m b_k^j u^k \frac{\partial}{\partial x^j} (a_m^i x^m) \frac{\partial}{\partial x^i} f + S \\
&= \sum_j \sum_i \sum_k a_j^i b_k^j u^k \frac{\partial}{\partial x^i} f + S \\
&= \sum_i (ABu)^i \frac{\partial}{\partial x^i} f + S
\end{aligned}
\tag{4.24}
$$

Subtracting equation (4.23) from the last line above gives:

$$[L_U, L_X]f - [L_X, L_U]f = \sum_i (ABu)^i \frac{\partial}{\partial x^i} f \equiv L_{U(1)} f$$

The vector field corresponding to $L_{U(1)}$ will be denoted by

$$U^{(1)} = ABu$$

The notation anticipates defining $U^{(k)} = A^k Bu$, where A^k is the kth power of A. To be consistent, let $U^{(0)}$ replace U. It can be seen by repeating the procedure in equations (4.24) that $[L_{U(k)}, L_X] = L_{U(k+1)}$. Hence, one has that

$$[U^{(k)}, X] = U^{(k+1)} = A^{k+1} Bu$$

The Cayley-Hamilton theorem says that the nth power of an $n \times n$ matrix A is a linear combination of the lower powers of the matrix:

$$A^n = \sum_{i=1}^{n-1} \alpha^i A^i$$

By the Cayley-Hamilton theorem, the process of bracketing terminates with

$$[U^{(n-1)}, X] = U^{(n)} = \sum_{i=0}^{n-1} \alpha^i U^{(i)}$$

The Lie algebra has the following multiplication table:

	$X]$	$U^{(j)}]$
$[X,$	0	$-U^{(j+1)}$
$[U^{(k)},$	$U^{(k+1)}$	0

and every vector field Z can be written as the sum of vector fields $X, U^{(0)}, \cdots, U^{(n-1)}$.

That every vector field can be written in these terms is related to the notion of controllability of control systems. In fact, as was recalled in chapter 2, a well known criterion for the complete controllability of the linear constant coefficient system $\dot{x} = Ax + Bu$ is that the matrix, whose columns are obtained from the columns of $A^k B$, where A is $n \times n$, have rank n:

$$\text{rank } (B, AB, \cdots, A^{n-1}B) = n$$

This is one example showing that the theory of the controllability of systems is related to the dimension of the Lie algebra generated by the families of vector fields. The literature is rich in regard to the connection with nonlinear systems (see, e.g., [10]). The Lie algebra of all vector fields on a manifold seems to be a very difficult object to study. There are many mathematical questions involving this algebra which will probably not be answered in the near future. However, in chapter 6 we show that when the manifold is a Lie group there is a subalgebra which is intimately related to the Lie group structure. We will pause in our considerations of the calculus, and in the meantime, in the next chapter, consider purely algebraic operations.

Chapter 5

EXTERIOR ALGEBRA

The purpose of this chapter is to introduce the operations of exterior algebra and to illustrate their uses. This algebra pervades present day differential geometry. Indeed, it appears to be an essential tool of the calculus. It also proves to be useful in its simple algebraic aspects apart from its use in calculus. It can be as notationally clear and as operationally direct as the usual Gibbsian vector algebra. It is superior to that algebra in two respects: it is conceptually more consistent that the Gibbsian scheme; and it is at home in any finite number of dimensions.

Forms and vectors are basic objects in the algebra. The two are quite different in their geometry in calculus, where vectors belong to tangent spaces and forms belong to cotangent spaces. Viewed purely algebraically, however, these differences are harder to see. It will smooth the transition to their use in differential geometry, though, if we keep vectors and forms distinct at least in name. In this chapter, most of the time we will be speaking of forms. Additional information on the algebra can be found in [11].

5.1   Addition of Forms

The forms in an n-dimensional space come in $n+1$ classes or orders. If $n = 3$, for example, the four orders run from the zero'th through the third. The zero'th order forms are simply constants, for example, a real number. The 1-form a can be written in components as $a = a, 1 + a_2 2 + a_3 3$. The numerals stand as *unit forms* exactly as in the usual vector notations. If a is a 2-form, then it can be written as $a = a_{12} 12 + a_{31} 31 + a_{23} 23$, or, as $a = a_{12} 1 \wedge 2 + a_{31} 3 \wedge 1 + a_{23} 2 \wedge 3$. If it is a 3-form, then, like the 0-form, it has only one component: $a = a_{123} 1 \wedge 2 \wedge 3$. Each order forms a vector space often denoted by Λ. If a is a p-form, then $a \in \Lambda^p$. By calling Λ^p a vector space, we mean that its elements can be multiplied by real numbers and added together with results that also are members of the space. The addition is associative and commutative, and the multiplication is distributive. That is, if $a, b, c \in \Lambda^p$ and $\alpha, \beta \in R$ then:

$$a + (b + c) = (a + b) + c = c + a + b$$

$$\alpha(a + b) = \alpha a + \alpha b = a\alpha + b\alpha$$
$$a(\alpha + \beta) = \alpha a + \beta a$$

Finally, there is an identity element, 0, in Λ^p such that $a + 0 = a$ for any element $a \in \Lambda^p$.

	1	2	3
$1\wedge$	0	$1\wedge 2$	$1\wedge 3$
$2\wedge$	$-1\wedge 2$	0	$2\wedge 3$
$3\wedge$	$-1\wedge 3$	$2\wedge 3$	0

	1	2	3
$1\wedge 2\wedge$	0	0	$1\wedge 2\wedge 3$
$2\wedge 3\wedge$	$1\wedge 2\wedge 3$	0	0
$3\wedge 1\wedge$	0	$1\wedge 2\wedge 3$	0

Figure 5.1: Multiplication in Λ^2

5.2 The Wedge Product

There is also a multiplication of forms. Its symbol is the wedge $\wedge$. It is antisymmetric, so that for unit forms $1, 2, 1\wedge 2 = -2\wedge 1$. Thus the order in which the elements stand in a terms is important and not generally commutable. The multiplication table for unit forms in Λ^1 of a three-dimensional space and is shown in the Figure 5.1.

The product terms belong to the vector space Λ^2. They, in turn, can have the multiplication table shown in Figure 5.1. This is as high as one goes in a 3-dimensional space, for any further multiplication would require a term that repeats one of the digits 1, 2 or 3. The antisymmetry of the multiplication prohibits the repetition. The properties of the products are summarized in Figure 5.2 where a, b, and c are forms with $a \in \Lambda^p$ and $b \in \Lambda^q$, and α and β are real numbers. It comes from the first line of the figure that if p is odd, $a \wedge a = 0$. This conclusion gives the zeros in the tables in the previous figures. It should be noted

$$a \wedge b = (-1)^{pq} b \wedge a \in \Lambda^{p+q} \text{ for } p + q \leq n$$
$$a \wedge (b \wedge c) = (a \wedge b) \wedge c = a \wedge b \wedge c$$
$$\alpha a \wedge \beta b = \alpha \beta a \wedge b = a \wedge b \alpha \beta$$

Figure 5.2: Multiplication of Forms

and understood also, that the second line of Figure 5.2 is subject to the inequality indicated in the first line.

The examples of multiplication in 3 dimensions show that Λ^1 and Λ^2 each have 3 independent unit forms and thus have dimension 3. Figure 5.1 shows Λ^3 with only one dimension. Similarly, Λ^0 has but one. In an n-dimensional space, Λ^p has as many independent forms as there are combinations of n things taken p at a time:

$$\dim \Lambda^p = C_p^n = \frac{n!}{(p!)(n-p)!}$$

Thus, if $n = 2$, $\dim \Lambda^0 = 1 = \dim \Lambda^2$; $\dim \Lambda^1 = 2$. (That $\dim \Lambda^1 = \dim \Lambda^{n-1} = n$ only for $n = 1, 2$, or 3 limits the utility of the Gibbsian vector algebra). The number of independent forms available together with the antisymmetry of the multiplication gives the dimensionality of Λ^p to be C_p^n. It can also be deduced from the factorial expression that $\dim \Lambda^{n-p} = \dim \Lambda^p$.

Some examples of the use of the wedge product are shown in Figure 5.3 for 1-forms in 3-space. No apologies are made for writing everything out in coordinate form since eventually in a computation one must come to a particular choice of coordinates. The beauty of the algebra, of course like that of the Gibbsian algebra, is that general conclusions can be drawn from manipulating coordinate-free expressions.

The first example shows 12 used as shorthand for $1 \wedge 2$. It can also be noted that 12, 23, and 31 are chosen as unit forms. This choice is an expression of the convention of choosing the order of digits as positive when the order is taken from the sequence 1,2,3,1,2, $\cdots$. It is the cyclic continuation of the natural order 1,2,3 and shows all even permutations of that order. This convention is also used in the second example, which additionally illustrates associativity of the multiplication as well as the fact that $\dim \Lambda^3 = 1$.

The complexity of choice of order for the factors of the unit forms increases with the dimensionality of the space. The awkwardness can be seen in choosing the order of factors for a 2-form in 4 dimensions. Once it has been chosen to continue the order selected for 2-forms in three dimensions as 12,23, and 31, then the choice is only between 14 or 41 and 24 or 42. One basis for selecting the order for these pairs is to following the hint given by the second example in Figure 5.3 and multiply by another form to land the product in $\Lambda^n = \Lambda^4$. Since there is only one term in Λ^4, all expressions of the unit 4-form that are even permutations of the natural order can be taken as positive; the others, negative.

The reader has probably compared the first example of Figure 5.3 with the result obtained from the expansion of the Gibbsian cross product $a \times b$, and the second example with the *triple scalar product* $c \cdot a \times b$ or $c \times a \cdot b$. The coefficients are the same in both algebras. The cross product

$$
\begin{aligned}
a \wedge b \;=\;& (a_1 1 + a_2 2 + a_3 3) \wedge (b_1 1 + b_2 2 + b_3 3) \\
=\;& (a_1 b_2 - a_2 b_1) 1 \wedge 2 + (a_2 b_3 - a_3 b_2) 2 \wedge 3 + \\
& (a_3 b_1 - a_1 b_3) 3 \wedge 1 \\
=\;& (a_1 b_2 - a_2 b_1) 12 + (a_2 b_3 - a_3 b_2) 23 + \\
& (a_3 b_1 - a_1 b_3) 31
\end{aligned}
$$

$$
\begin{aligned}
c \wedge a \wedge b \;=\;& (c_1 1 + c_2 2 + c_3 3) \wedge a \wedge b \\
=\;& (c_1 (a_2 b_3 - a_3 b_2) 123 + c_2 (a_3 b_1 - a_1 b_3) 231 + \\
& c_3 (a_1 b_2 - a_2 b_1) 312 \\
=\;& [c_1 (a_2 b_3 - a_3 b_2) + c_2 (a_3 b_1 - a_1 b_3) + \\
& c_3 (a_1 b_2 - a_2 b_1)] 123 \\
=\;& [(c_2 a_3 - c_3 a_2) b_1 + (c_3 a_1 - c_1 a_3) b_2 + \\
& (c_1 a_2 - c_2 a_1) b_3] 123 \\
=\;& c \wedge a \wedge (b_1 1 + b_2 2 + b_3 3)
\end{aligned}
$$

Figure 5.3: Examples of Wedge Product

$$V \lrcorner a \in \Lambda^{p-1}(: V \lrcorner \alpha = 0) \tag{5.1}$$

$$\beta V \lrcorner \alpha a = \alpha \beta V \lrcorner a \tag{5.2}$$

$$V \lrcorner (\alpha a + \beta b) = \alpha V \lrcorner a + \beta V \lrcorner b \tag{5.3}$$

$$(\alpha U + \beta V) \lrcorner a = \alpha U \lrcorner a + \beta V \lrcorner b \tag{5.4}$$

$$V \lrcorner (a \wedge c) = (V \lrcorner a) \wedge c + (-1)^p a \wedge (V \lrcorner c) \tag{5.5}$$

Figure 5.4: Contraction

object appears to be in the same space as its factors, however, whereas the result of the wedge operation is clearly in a different space, namely in Λ^2. Again, whereas the triple scalar product appears to be a real number, its exterior counterpart shows otherwise. That the objects of exterior algebra are exactly what they appear to be and those of the Gibbsian algebra are not, is seen more clearly when we investigate how they behave under linear transformations.

5.3 Contraction of Forms; Vectors

The wedge product of a p-form a and a 1-form b is a $(p+1)$-form: $a \in \Lambda^p, b \in \Lambda^1; a \wedge b = c \in \Lambda^{p+1}$ (if $p + 1 \leq n$). A

	1	2	3
$\bar{1}\,\lrcorner$	(1)	0	0
$\bar{2}\,\lrcorner$	0	(1)	0
$\bar{3}\,\lrcorner$	0	0	(1)

Figure 5.5: Contraction of Vectors and Their Duals

product can also be defined that reduces the p-order of a form. The operation is often symbolized by $\lrcorner$, called the *angle*. The angle product represents the action of a vector contracting the order of a p-form in n-dimensional space. That is to say, vectors contract forms. Let: α and β be real numbers; a and b be p-forms and c a q-form such that $p + q \leq n$, the dimension of the space; U and V be vectors. Then the vector action is shown in Figure 5.4.

The parenthetical remark in statement (5.1) tries to say that the angle action on, or the contraction of, a real number is defined to be zero, just as the wedge product leading to a $p + q > n$ is zero. Statements (5.2), (5.3) and (5.4) show that the action is linear in both variables over the operations of a vector space. Statement (5.5) shows how the action distributes itself over the wedge product.

Statement (5.1) also says that a vector maps a 1-form into a real number. The result is to have all the properties of the usual inner product, except for symmetry in the formalism. That is, here the *vector angle product* is written as a vector operating on the form. As an inner product, then, $V \lrcorner a \rightarrow \mathbf{R}$ defines the unique V dual to a. The operation can be visualized if V is *coordinatized* and the angle operation is illustrated by the dual unit vectors. Thus, let $V = v_1\bar{1} + v_2\bar{2} + v_3\bar{3}$ and $a = a_1 1 + a_2 2 + a_3 3$. The unit vectors

	12	23	31
$\bar{1}\,\lrcorner$	2	0	−3
$\bar{2}\,\lrcorner$	−1	3	0
$\bar{3}\,\lrcorner$	0	−2	1

Figure 5.6: Contraction of Dual Vectors and 2-Forms

$\bar{1}, \bar{2}$ and $\bar{3}$, distinguished by the bar above the number, are dual to the set 1,2 and 3. Then $V \lrcorner a = v_1 a_1 + v_2 a_2 + v_3 a_3$. The multiplication table is shown in Figure 5.5. The (1) in the figure is the real number one, written in parentheses to distinguish it from a unit form. The signs in the multiplication table in the figure illustrate the commutation rule in statement (5.5) of Figure 5.4. The table also demonstrates how the general commutation law can be turned into the handy computational rule: *always contract over the first factor of a wedge product*; or, bring the factor being contracted to the front of the form in the wedge products. The rule is illustrated in the table in Figure 5.6.

Let c be a 2-form in 3-space. Then $V \lrcorner c$ is a 1-form and can be acted on by another vector $U : U \lrcorner V \lrcorner c \in \Lambda^0$. Figure 5.7 gives explicit examples of these products, using the tables in the previous figures.

The second example in Figure 5.7 shows an antisymmetry redolent of the wedge product: $(U \lrcorner V \lrcorner) = -(V \lrcorner U \lrcorner)$. The reader can show himself that the vectors enjoy an algebra entirely analogous to that furnished for forms. This fact is good to keep in mind, but it does not appear to lead to much that is useful in the vector calculus.

A final observation can be made on the same example.

$$V \lrcorner c \;=\; (v_1\bar{1} + v_2\bar{2} + v_3\bar{3}) \lrcorner (c_3 12 + c_2 31 + c_1 23)$$
$$=\; v_1 c_3 2 - v_1 c_2 3 - v_2 c_3 1 + v_2 c_1 3 + v_3 c_2 1 - v_3 c_1 2$$
$$=\; (v_3 c_2 - v_2 c_3)1 + (v_1 c_3 - v_3 c_1)2 +$$
$$(v_2 c_1 - v_1 c_2)3 \in \Lambda^1$$

$$U \lrcorner V \lrcorner c \;=\; (u_1\bar{1} + u_2\bar{2} + u_3)\bar{3} \lrcorner V \lrcorner c$$
$$=\; u_1(v_3 c_2 - v_2 c_3) + u_2(v_1 c_3 - v_3 c_1) +$$
$$u_3(v_2 c_1 - v_1 c_2)$$
$$=\; (v_2 u_3 - v_3 u_2)c_1 + (v_3 u_1 - v_1 u_3)c_2 +$$
$$(v_1 u_2 - v_2 u_1)c_3 \in \Lambda^0$$

Figure 5.7: Higher Order Products

Note that $U \lrcorner V \lrcorner c$ is a real number. Thus $(U \lrcorner V)$ is dual to c and defines an inner product of a 2-vector and a 2-form. The reader will probably immediately investigate whether or not one can talk of a 3-vector, for example $w \lrcorner u \lrcorner v$, acting on he unit 3-form 123. One finds an expression analogous to the product $c \wedge a \wedge b$ shown in Figure 5.3, except that the result of acting on the 3-form is a real number. Despite the importance of these matters of contraction, lying as they do at the heart of integration, they will not be discussed further here. (They are nicely discussed in Flander's book [11].)

5.4 Equation of a Plane

A simple example of the use of the wedge product gives the non-parametric form of the equation for a plane. The parametric equation for a plane in 3-space is given by the expression

$$r = r_0 + \alpha a + \beta b$$

with r_0, a and b fixed 1-forms and α and β parameters that can have any real number as a value. Let $\Delta r = r - r_0$, so that the expression becomes $\Delta r = \alpha a + \beta b$. Multiplying both sides of this equation by $a \wedge b$ eliminates the parameters α and β:

$$
\begin{aligned}
\Delta r \wedge a \wedge b &= (\Delta x 1 + \Delta y 2 + \Delta z 3) \wedge [(a_2 b_3 - a_3 b_2)23 + \\
&\quad (a_3 b_1 - a_1 b_3)31 + (a_1 b_2 - a_2 b_1)12] \\
&= 0
\end{aligned}
$$

When the indicated multiplication has been carried out, it

will be seen to express the expansion of the determinant

$$\begin{vmatrix} \Delta x & a_1 & b_1 \\ \Delta y & a_2 & b_2 \\ \Delta z & a_3 & b_3 \end{vmatrix}$$

as the coefficient of the 3-form 123. The form is interpreted geometrically as a unit of volume with an orientation given by the order chosen by the order 1,2,3.

Expansion of the wedge product or of the determinant gives the familiar equation of the plane:

$$Ax + By + Cz = D = Ax_0 + By_0 + Cz_0$$

where the coefficients A, B and C are the value of the determinants:

$$A = \begin{vmatrix} a_2 & b_2 \\ a_3 & b_3 \end{vmatrix}$$

$$B = -\begin{vmatrix} a_1 & b_1 \\ a_3 & b_3 \end{vmatrix}$$

$$C = \begin{vmatrix} a_1 & b_1 \\ a_2 & b_2 \end{vmatrix}$$

Their determinantal form suggests that the unit 2-forms can be interpreted geometrically as oriented areas; that is, areas with a sign prescribed. The procedure for eliminating parameters to get the nonparametric form of the equation of a plane, and the interpretation of signed volumes and areas readily generalize to any finite number of dimensions.

What we call forms in this section could as easily have been called vectors, and the analogous operations would

$$a : (a_1, a_2, a_3) \rightsquigarrow \begin{bmatrix} a_1 \\ a_2 \\ a_3 \end{bmatrix} ; b : (b_1, b_2, b_3) \rightsquigarrow \begin{bmatrix} b_1 \\ b_2 \\ b_3 \end{bmatrix}$$

Figure 5.8: Determinant Format

carry through. Signed areas and volumes perhaps should be considered as generalized vectors even here, because they are the things forms get integrated over or along. From the point of view of the algebra, it doesn't matter much how they are referred to, as long as quantities of both kinds are not being handled at once.

5.5 Use of Determinants

The exterior algebra is often called *Grassman* algebra after the person who developed its properties in the 1840's. Klein [12] says that Grassman utilized determinants to discuss the algebra of finite figures such as line segments (as distinguished from lines of infinite lengths), and triangles or parallelograms that bound a finite and oriented area. Grassmann pointed out that all the sub-determinants that can be formed from the determinant of a finite volume have a geometric interpretation that permits an algebra of the figures. This section will use these ideas in illustrating another representation of forms (or vectors).

We have been representing forms as sets of real numbers, using these numbers or components to size the forms along orthogonal unit forms. Another format, useful in some computations and in visualizing some expressions of

$$a \wedge b : \begin{bmatrix} a_1 \\ a_2 \\ a_3 \end{bmatrix} \wedge \begin{bmatrix} b_1 \\ b_2 \\ b_3 \end{bmatrix} \rightarrow \begin{bmatrix} a_1 & b_1 \\ a_2 & b_2 \\ a_3 & b_3 \end{bmatrix} \rightsquigarrow \begin{bmatrix} \begin{vmatrix} a_1 & b_1 \\ a_2 & b_2 \end{vmatrix} \\ \begin{vmatrix} a_1 & b_1 \\ a_3 & b_3 \end{vmatrix} \\ \begin{vmatrix} a_2 & b_2 \\ a_3 & b_3 \end{vmatrix} \end{bmatrix}$$

Figure 5.9: Determinant Representation of Wedge Product

transformation, can also be adopted. It looks like a matrix format but is intended to signify determinants. In this format, the 1-forms a and b in 3-space appear as in Figure 5.8. One can manipulate these representations in a consistent way to express the algebra.

Figure 5.9 shows the wedge product of the two columns a and b as a 3×2 matrix. This matrix stands for the set of all 2×2 determinants that can be obtained directly from the matrix. These determinants are shown on the right in the figure. The correspondence with determinants was foreshadowed by the expansion of $\Delta r \wedge a \wedge b$ in the previous section.

For more examples, consider the *unit forms* in Figure 5.10. The first example gives a zero 2-form, as it should. The second gives a unit 2-form.

The third example, shows the unit 3-form. The importance of order in the wedge product is reflected in the usual definition of the determinants on the right in the figure. The

$$
\begin{bmatrix} 1 \\ 0 \\ 0 \end{bmatrix} \wedge \begin{bmatrix} 1 \\ 0 \\ 0 \end{bmatrix} \rightarrow \begin{bmatrix} 1 & 1 \\ 0 & 0 \\ 0 & 0 \end{bmatrix} \rightsquigarrow \begin{bmatrix} \begin{vmatrix} 1 & 1 \\ 0 & 0 \end{vmatrix} \\ \begin{vmatrix} 1 & 1 \\ 0 & 0 \end{vmatrix} \\ \begin{vmatrix} 0 & 0 \\ 0 & 0 \end{vmatrix} \end{bmatrix}
$$

$$
\begin{bmatrix} 1 \\ 0 \\ 0 \end{bmatrix} \wedge \begin{bmatrix} 0 \\ 1 \\ 0 \end{bmatrix} \rightarrow \begin{bmatrix} 1 & 0 \\ 0 & 1 \\ 0 & 0 \end{bmatrix} \rightsquigarrow \begin{bmatrix} \begin{vmatrix} 1 & 0 \\ 0 & 1 \end{vmatrix} \\ \begin{vmatrix} 1 & 0 \\ 0 & 0 \end{vmatrix} \\ \begin{vmatrix} 0 & 1 \\ 0 & 0 \end{vmatrix} \end{bmatrix}
$$

$$
\begin{bmatrix} 0 \\ 1 \\ 0 \end{bmatrix} \wedge \begin{bmatrix} 0 \\ 0 \\ 1 \end{bmatrix} \wedge \begin{bmatrix} 1 \\ 0 \\ 0 \end{bmatrix} \rightarrow \begin{bmatrix} 0 & 0 & 1 \\ 1 & 0 & 0 \\ 0 & 1 & 0 \end{bmatrix} \rightsquigarrow \begin{vmatrix} 0 & 0 & 1 \\ 1 & 0 & 0 \\ 0 & 1 & 0 \end{vmatrix}
$$

Figure 5.10: Wedge Product of Unit Vectors

curved arrow in this and the previous figures is meant to signify a correspondence between the items it joins.

5.6　Solution of Linear Equations

Using the determinantal representation to solve linear algebraic equations will now be seen to put *Cramer's rule* into a consistent setting. Express the equation $Ax = B$ in the following way:　$A_1x_1 + A_2x_2 + A_3x_3 = B$ with A_i given as 1-forms with the representation

$$
\begin{bmatrix} a_{11} \\ a_{22} \\ a_{33} \end{bmatrix} x_1 + \begin{bmatrix} a_{12} \\ a_{22} \\ a_{32} \end{bmatrix} x_2 + \begin{bmatrix} a_{13} \\ a_{23} \\ a_{33} \end{bmatrix} x_3 = \begin{bmatrix} b_1 \\ b_2 \\ b_3 \end{bmatrix} \qquad (5.6)
$$

Then, as in forming the non-parametric equation of a plane discussed in Section 5.4, multiply equation (5.6) by $A_2 \wedge A_3$:

$$
\begin{aligned}
A_1 \wedge A_2 \wedge A_3 x_1 &+ A_2 \wedge A_2 \wedge A_3 x_2 + A_3 \wedge A_2 \wedge A_3 x_3 \\
&= B \wedge A_2 \wedge A_3 \\
&= A_1 \wedge A_2 \wedge A_3 x_1 \\
&= B \wedge A_2 \wedge A_3
\end{aligned}
$$

since $A_2 \wedge A_2 \wedge A_3 = A_3 \wedge A_2 \wedge A_3 = 0$. For x_2, either multiply by $A_1 \wedge A_3$ with due regard for signs, or pre-multiply by A_1 and postmultiply by A_3 to give

$$
A_1 \wedge A_2 \wedge A_3 x_2 = A_1 \wedge B \wedge A_3 \qquad (5.7)
$$

$$\xi_1\eta_2 - \xi_2\eta_1 = \frac{B_1 \wedge B_2 \wedge A_3}{A_1 \wedge A_2 \wedge A_3}$$

$$\xi_1\eta_3 - \xi_3\eta_1 = \frac{B_1 \wedge B_2 \wedge A_2}{A_1 \wedge A_3 \wedge A_2}$$

$$\xi_2\eta_3 - \xi_3\eta_2 = \frac{B_1 \wedge B_2 \wedge A_1}{A_2 \wedge A_3 \wedge A_1}$$

Figure 5.11: Cramer's Rule

With $A_1 \wedge A_2 \wedge A_3, B \wedge A_2 \wedge A_3$ and $A_1 \wedge B \wedge A_3$ in the determinantal forms:

$$\begin{vmatrix} a_{11} & a_{12} & a_{13} \\ a_{21} & a_{22} & a_{23} \\ a_{31} & a_{32} & a_{33} \end{vmatrix}, \begin{vmatrix} b_1 & a_{12} & a_{13} \\ b_2 & a_{22} & a_{23} \\ b_2 & a_{32} & a_{33} \end{vmatrix}, \begin{vmatrix} a_{11} & b_1 & a_{13} \\ a_{21} & b_2 & a_{23} \\ a_{31} & b_3 & a_{33} \end{vmatrix}$$

the correspondence with Cramer's rule is easily recognized.

The same procedure applies in the case of 2-forms. With $x = \xi \wedge \eta$, the equation $Ax = B$ can be written as:

$$\begin{bmatrix} a_{11} & a_{12} & a_{13} \\ a_{21} & a_{22} & a_{23} \\ a_{31} & a_{32} & a_{33} \end{bmatrix} \xi \wedge \eta = \begin{bmatrix} b_{11} & b_{12} \\ b_{21} & b_{22} \\ b_{31} & b_{32} \end{bmatrix} \tag{5.8}$$

Equation (5.8) can be written to show more explicitly that it is an equation of 2-forms:

$$\begin{bmatrix} a_{11} & a_{12} \\ a_{21} & a_{22} \\ a_{31} & a_{32} \end{bmatrix} (\xi_1\eta_2 - \xi_2\eta_1) + \begin{bmatrix} a_{11} & a_{13} \\ a_{21} & a_{23} \\ a_{31} & a_{33} \end{bmatrix} (\xi_1\eta_3 - \xi_3\eta_1)$$

$$+ \begin{bmatrix} a_{12} & a_{13} \\ a_{22} & a_{23} \\ a_{32} & a_{33} \end{bmatrix} (\xi_2\eta_3 - \xi_3\eta_2) = \begin{bmatrix} b_{11} & b_{12} \\ b_{21} & b_{22} \\ b_{31} & b_{32} \end{bmatrix} \qquad (5.9)$$

Even in this form the equation is symbolic rather than explicit because each of the matrices in equation (5.9) stands for a set of determinants as shown in Figure 5.9. Following the form in equation (5.6), one can write equation (5.8) as:

$$A_1 \wedge A_2(\xi_1\eta_2 - \xi_2\eta_1) + A_1 \wedge A_3(\xi_1\eta_3 - \xi_3\eta_1)$$
$$+ A_2 \wedge A_3(\xi_2\eta_3 - \xi_3\eta_2) = B_1 \wedge B_2 \qquad (5.10)$$

The procedure used to solve equation (5.6) suggests that equation (5.10) be multiplied in turn on the right by A_3, A_2 and then A_1. The solution that results is shown in Figure 5.11. The procedure corresponds to an application of Cramer's rule and is easily recognized. Furthermore, the procedure is not limited to 1- and 2-forms nor to 3 dimensions.

5.7   Linear Transformations

We remarked in the introduction to this chapter that exterior algebra is superior to the Gibbsian vector algebra in that it is conceptually more consistent and that it is at home in any finite number of dimensions. We have called attention to the truth of the second aspect before, and the reader will see it illustrated again here. The truth of the first aspect will be particularly well illustrated here with linear transformations, where the consistence of the formalism of exterior algebra contrasts with the unexpected rule of transformation of cross products of the common vector algebra.

This section is to illustrate that the form of the equation (5.6) holds in expressing a linear transformation, regardless of the dimension of the space or the order of the objects x and B. Although the illustration is limited to 3-space and to 1- and 2-forms, the point illustrated clearly holds beyond these limits.

Let x, y, ξ, and η be 1-forms. Let $x = A\xi$ and $y = A\eta$. In detail,

$$x = A\xi : \begin{bmatrix} x_1 \\ x_2 \\ x_3 \end{bmatrix} = \begin{bmatrix} a_{11} & a_{12} & a_{13} \\ a_{21} & a_{22} & a_{23} \\ a_{31} & a_{32} & a_{33} \end{bmatrix} \begin{bmatrix} \xi_1 \\ \xi_2 \\ \xi_3 \end{bmatrix} \qquad (5.11)$$

with a similar expression for $y = A\eta$. Then transformation of the wedge product $x \wedge y$ has the same form:

$$x \wedge y = A\xi \wedge \eta : \begin{bmatrix} x_1 y_2 - x_2 y_1 \\ x_1 y_3 - x_2 y_1 \\ x_2 y_3 - x_3 y_2 \end{bmatrix} =$$
$$\begin{bmatrix} a_{11} & a_{12} & a_{13} \\ a_{21} & a_{22} & a_{23} \\ a_{31} & a_{32} & a_{33} \end{bmatrix} \begin{bmatrix} \xi_1 \eta_2 - \xi_2 \eta_1 \\ \xi_1 \eta_3 - \xi_3 \eta_1 \\ \xi_2 \eta_3 - \xi_3 \eta_2 \end{bmatrix} \qquad (5.12)$$

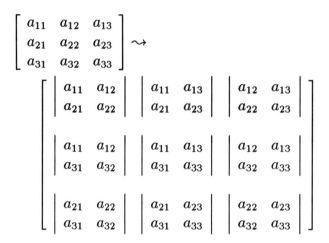

Figure 5.12: Linear Transformations

However, because the transformation is of 2-forms, the matrix A in equation (5.12) must be understood to have as components all 2×2 determinants that can be obtained from the matrix A. the identification shown in Figure 5.12 generalizes that shown in Figure 5.9.

How equation (5.12) together with Figure 5.12 came about can be seen by multiplying out the product of the real numbers involved. The products for $x_1 y_2$ and $x_2 y_1$ are written out for illustration in Figure 5.13. The rest of the elements in Figure 5.12 can be obtained similarly. It is the antisymmetry of the wedge product which gives the consistent result that the components of the matrix A are p-forms if the transformed quantities are the products of p 1-forms, each transformed in the same way.

$$
\begin{aligned}
x_1 y_2 &= (a_{11}\xi_1 + a_{12}\xi_2 + a_{13}\xi_3) \times \\
&\quad (a_{21}\eta_1 + a_{22}\eta_2 + a_{23}\eta_3) \\
&= a_{11}a_{21}\xi_1\eta_1 + a_{11}a_{22}\xi_1\eta_2 + a_{11}a_{23}\xi_1\eta_3 \\
&\quad + a_{12}a_{21}\xi_2\eta_1 + a_{12}a_{22}\xi_2\eta_2 + a_{12}a_{23}\xi_2\eta_3 \\
&\quad + a_{13}a_{21}\xi_3\eta_1 + a_{13}a_{22}\xi_3\eta_2 + a_{13}a_{23}\xi_3\eta_3
\end{aligned}
$$

$$
\begin{aligned}
x_2 y_1 &= (a_{21}\xi_1 + a_{22}\xi_2 + a_{23}\xi_3) \times \\
&\quad (a_{11}\eta_1 + a_{12}\eta_2 + a_{13}\eta_3) \\
&= a_{21}a_{11}\xi_1\eta_1 + a_{21}a_{12}\xi_1\eta_2 + a_{21}a_{13}\xi_1\eta_3 \\
&\quad + a_{22}a_{11}\xi_2\eta_1 + a_{22}a_{12}\xi_2\eta_2 + a_{22}a_{13}\xi_2\eta_3 \\
&\quad + a_{23}a_{11}\xi_3\eta_1 + a_{23}a_{12}\xi_3\eta_2 + a_{23}a_{13}\xi_3\eta_3
\end{aligned}
$$

$$
\begin{aligned}
x_1 y_2 - x_2 y_1 &= (a_{11}a_{22} - a_{21}a_{12})(\xi_1\eta_2 - \xi_2\eta_1) \\
&\quad + (a_{11}a_{23} - a_{21}a_{13})(\xi_1\eta_3 - \xi_3\eta_1) \\
&\quad + (a_{12}a_{23} - a_{22}a_{13})(\xi_2\eta_3 - \xi_3\eta_2) \\
&= \left[\begin{array}{cc|cc|cc}
a_{11} & a_{12} & a_{11} & a_{13} & a_{12} & a_{13} \\
a_{21} & a_{22} & a_{21} & a_{23} & a_{22} & a_{23}
\end{array}\right] \times \\
&\quad \begin{bmatrix}
\xi_1\eta_2 - \xi_2\eta_1 \\
\xi_1\eta_3 - \xi_3\eta_1 \\
\xi_2\eta_3 - \xi_3\eta_2
\end{bmatrix}
\end{aligned}
$$

Figure 5.13: An Explicit Calculation

Chapter 6

LIE GROUPS AND ACTIONS

The connection between Lie algebras and Lie groups is seen in this chapter to be the same as the relationship between a linear differential equation and its solution. The connection is discussed after what is meant by a group, by a Lie group, and by the action of a group have been clarified.

6.1   Lie Groups

Since the idea of a group is a purely abstract algebraic idea, the definition of a group should involve only a set of elements and some algebraic relations between them. A group, then, is a set of elements, like a set of matrices, any pair of which can operate together to give another element that is also in the same set: (the product of two $n \times n$ matrices is an $n \times n$ matrix). Matrix multiplication is the operation for this example. When the integers are viewed as forming a group, addition is the group operation. Besides having an operation between elements that yields another element of a set, a group requires the following conditions: to each element of the group there corresponds an inverse that is also in the group, one element in the group acts as an identity element (unity, in the case of multiplication; zero, in the case of addition), and, finally, the operation is associative.

Among matrices, for example, consider the unitary matrices discussed in chapter 2 (unitary matrices who elements are real are orthogonal matrices). The set of all $n \times n$ unity matrices, $U(n)$, forms a group with the usual matrix product as the operation: if A and B belong to $U(n)$, so does AB. Since for every A there is an A^T, its transpose, such that $A^T A = AA^T = e$, A^T is the inverse of A. Also, the matrix e (or I) is the identity element belonging to $U(n)$, and $AI = A$ for every $A \in U(n)$.

The discussion in chapter 2 showed that the group $U(n)$ is a manifold, and that this dimension is $(1/2)n(n-1)$. The product of two unity matrices is a unitary matrix, and chapter 2 showed that this operation is a manifold map. Furthermore, since the elements of the product consist of sums of products of the elements of the two factor matrices and which are real or complex numbers, they are C^∞ functions (even analytic functions, that is, expressible as a Taylor's series). The above properties characterize what are called Lie groups (or 'continuous transformation groups', in the older literature). Formally, a Lie group is a group that is a manifold and whose group operation yields C^∞ manifold functions and is associative. Although Lie group elements need not be written as matrices, they will be thought of in that way in the sequel.

To illustrate the notation, which follows the usual group notation, let g, h, and k be elements of the group. Denoting the group operation by '$\cdot$', gives $\cdot(,) : G \times G \to G$; thus, $\cdot(g,h) = gh$. Associativity gives that $\cdot(k, \cdot(g,h)) = \cdot(\cdot(k,g), h)$. For the inverse, $()^{-1} : G \to G$; thus $(g)^{-1} = g^{-1}$. The $\cdot(,)$ and $()^{-1}$ are C^∞ functions. The identity element, e, is a distinguished element of the group such that $eg = ge = g$ for any and all $g \in G$.

6.2   Group Action

The elements of a group are not restricted to operating among themselves. Their importance, in fact, comes from their operating on objects that don't belong to the same set. Thus, the importance of matrices is not that they may belong to a group, but that they operate on vectors to form new vectors in old frames or old vectors in new frames. The action of a group refers to members of a group operating on nonmembers.

Let M be a manifold and G be a Lie group. Suppose that there is a C^∞ function $\tau : G \times M \to M$ (with $\tau(g, m_1) = m_2$, where m_1 and m_2 both belong to M) that has the following properties:

(i) for all $m \in M, \tau(e, m) = m$

(ii) $\tau(g, \tau(h, m)) = \tau(gh, m)$

We say in this case that the group G acts on M. Any group of $n \times n$ matrices, say the general n-dimensional linear group over the reals, $Gl(n, \mathbf{R})$, then, in operating on vectors in Euclidean space by the usual matrix multiplication, is said to act on $\mathbf{R}^n$.

Another example, with a somewhat more involved action, which is not linear, can be described in terms of linear fractional transformations. Let $\mathbf{R}^2$ be identified with $\mathbf{C}$. Let $Gl(2\mathbf{C})$ be the group of 2×2 matrices with complex elements. Define a map $\tau : Gl(2\mathbf{C}) \times \mathbf{C} \to \mathbf{C}$ as

$$\tau\left(\begin{bmatrix} a & b \\ c & d \end{bmatrix}, z\right) = \frac{az + b}{cz + d} = \omega \in \mathbf{C}$$

The mapping forms a group:

$$\tau\left(\begin{bmatrix} a_2 & b_2 \\ c_2 & d_2 \end{bmatrix}, \tau\begin{bmatrix} a_1 & b_1 \\ a_1 & b_1 \end{bmatrix}, z\right) =$$

$$\tau\left(\begin{bmatrix} a_2 & b_2 \\ c_2 & d_2 \end{bmatrix}\begin{bmatrix} a_1 & b_1 \\ c_1 & d_1 \end{bmatrix}, z\right)$$

$$= \frac{(a_2a_1 + b_2c_1)z + a_2b_1 + b_2d_1}{(c_2a_1 + d_2c_1)z + c_2b_1 + d_2d_1} \in \mathbf{C}$$

The group $Gl(2\mathbf{C})$ is said to act on $\mathbf{C}$ by the map τ. Now, the action is not well defined for $z = -d/c$. Hence, the particular maps must be considered to describe the action in local coordinates.

Exercise 8 *Define $\mathbf{R}$ acting on T^2 using the representation of T^2 as $S^1 \times S^1$. Choose the explicit representation of $S^1 = \{\exp i\theta : 0 \le \theta < 2\pi\}$. Define an action for each pair of positive real numbers (r, s) as*

$$(\alpha, e^{i\theta}, e^{i\gamma})) \to (e^{i(r\alpha + \theta)}, e^{i(s\alpha + \gamma)}).$$

For fixed values of θ and γ this creates a curve in T^2. When is the curve periodic? Prove that the curve is periodic iff $\frac{r}{s}$ is rational. Prove that if the curve is not periodic then the curve is dense in T^2. To do this, show that for any given point $(e^{i\theta}, e^{i\gamma})$ there is an α such that $(e^{i(r\alpha+\theta_0)}), e^{i(s\alpha+\gamma_0)})$ is close.

Remark: This exercise amounts to approximating (θ, γ) by $(r\alpha + \theta_0, s\alpha + \gamma_0)$ module 2π. Thus the problem of determining the denseness of the above curve is actually a problem in number theory. There are many beautiful applications of this theorem in the book on differential equations by Arnold. The curve with non-rational parameters is called the *irrational winding line* on the torus.

Exercise 9 *Interpret the winding line on the manifold $S = \{(x, y) : -1 \leq x < 1, -1 \leq y < 1\}$.*

Remark: This is one of a very general set of problems referred to as *Billiard Ball Problems*. In general, given a polygon, say for example a triangle such as in Figure 6.1 and a particle that rebounds from the sides in a physical manner, what are the conditions necessary for a dense path?

6.3   One-Parameter Subgroups

Consider a subgroup α of a Lie group, in which the elements of the subgroup are given in terms of a single parameter. One element is given by $\alpha(t)$. A neighboring element is given by $\alpha(t + h)$ if $t + h$ is close to t. One parameter subgroups of a Lie group are distinguished by the properties:

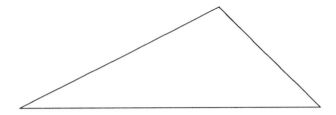

Figure 6.1: General Triangle

$$\alpha(t)\alpha(h) = \alpha(t + h) \qquad (6.1)$$
$$\alpha(0) = e \text{ the identity.}$$

One corollary of properties in (35) is that $\alpha^{-1}(t) = \alpha(-t)$.
Another corollary can be related to noting the basic group
property expressed in (35) makes a product of elements at
different values of the parameter correspond to an element
at the sum of the parameter values. A correspondence like
that is typical of exponential functions. How exponentials
get involved from these properties is seen from considering
derivatives:

$$\frac{d\alpha(t)}{dt} = \lim_{h \to 0} \frac{1}{h}[\alpha(t + h) - \alpha(t)]$$
$$= \lim_{h \to 0} \frac{1}{h}[(\alpha(h) - e)\alpha(t)]$$

$$= \dot{\alpha}(t) = A\alpha(t) \tag{6.2}$$

where A is the limit as $h \to 0$ of $(\alpha(h) - e)/h$. The limit exists because the group is a manifold whose coordinates are C^∞ functions. Equation (6.2) makes it easy to see why A is said to generate the subgroup, or is referred to as the infinitesimal generator of the subgroup.

If α and A are real numbers, equation (6.2) has the solution $\alpha(t)e^{At}\alpha(0) = e^{At}$, confirming the connection of properties (35) with exponentials. The same form holds if A is matrix, generating a matrix representation of the subgroup. In that case, the exponential function is understood to mean the series:

$$\exp(At) = I + At + A^2 t^2/2! + \cdots$$

These points can be illustrated by an example of a subgroup taken from the unitary group $U(2)$:

$$\alpha(t) = \begin{pmatrix} \cos t & \sin t \\ -\sin t & \cos t \end{pmatrix} \tag{6.3}$$

To illustrate (6.2):

$$\alpha(t)\alpha(h) =$$
$$\begin{pmatrix} \cos t \cos h - \sin t \sin h & \sin t \cos h + \cos t \sin h \\ -(\sin t \cos h + \cos t \sin h) & \cos t \cos h - \sin t \sin h \end{pmatrix}$$
$$= \begin{pmatrix} \cos(t+h) & \sin(t+h) \\ -\sin(t+h) & \cos(t+h) \end{pmatrix}$$

The derivative can be calculated directly or through examining the limit shown above and is found to be

$$\dot{\alpha} = \begin{pmatrix} 0 & 1 \\ -1 & 0 \end{pmatrix} \begin{pmatrix} \cos t & \sin t \\ -\sin t & \cos t \end{pmatrix} \qquad (6.4)$$

which gives for a representation of A the matrix:

$$A = \begin{pmatrix} 0 & 1 \\ -1 & 0 \end{pmatrix}$$

the solution of (6.4) should give (6.3), of course. To see that 'exponentiating' the generator: $A \rightarrow \exp(At)$ gives (6.3), note that

$$A^2 = \begin{pmatrix} -1 & 0 \\ 0 & -1 \end{pmatrix} = -I; \ A^3 = -A$$

Then

$$
\begin{aligned}
\exp(At) &= I + At + \frac{A^2 t^2}{2!} + \cdots \\
&= I \left(1 - \frac{t^2}{2!} + \frac{t^4}{4!} - \cdots \right) + \\
&\quad A \left(t - \frac{t^3}{3!} + \frac{t^5}{5!} - \cdots \right) \\
&= \begin{pmatrix} \cos t & \sin t \\ -\sin t & \cos t \end{pmatrix}
\end{aligned}
$$

as was expected.

Now, $\alpha(t)$ is a curve on the $U(2)$ manifold. To connect with earlier sections, note that it represents a homeomorphism of an interval of the real line $\mathbf{R}^1$ to $U(2)$. It is a manifold map $\mathbf{R}^1 \rightarrow U(2)$ and also is a coordinate function of the manifold. Indeed, since the dimension of $U(2)$ is one,

$\alpha(t)$ is the only coordinate function, or one realization of it. Furthermore, $\alpha(t)$ belongs to an equivalence class $[\alpha(t)]_e$. The tangent space at $t = 0$ is given explicitly as $I + At$, which is the tangent vector there. The vector field, X_A, is (I, A), and in the earlier notation, $\mathcal{X}_A = A$. The element of the tangent space is uniquely determined by A. This is seen by considering: $[\alpha]_e = [\beta]_e$, if and only if, $\dot{\alpha}(0) = \dot{\beta}(0)$, if and only if, $A\alpha(0) = B\beta(0)$, if and only if $A = B$.

It is thus clear that every point of a one-parameter Lie group satisfies a linear first-order differential equation. It thus gives rise to a vector field which is an infinitesimal generator for the subgroup. The converse is also true: every vector field acts as an infinitesimal generator of a one-parameter Lie group, the exponential form meaning the series expansion. The interpretation in the tangent space is simple: $X_A\alpha(0) = (I + At)\alpha(0)$.

Chapter 4 shows that a vector field is not only a vector space, but also an algebra. If $\mathcal{X}_A = A$, then $\mathcal{X}_{\rho A} = \rho A$, for ρ a real number: $\mathcal{X}_{A+B} = A + B$; and similarly, for $\mathcal{X}_{[A,B]}$, the commutator of A and B, namely, $AB - BA$. The product AB, however, does not generally belong to a vector field.

As $\rho A, A + B$, and $(AB - BA)$ are vector fields, they should be generators of Lie groups. For the first,

$$\dot{\gamma}(t) = \rho A\gamma(t) \to \gamma(t) = \exp(\rho At)\gamma(0) = \alpha(\rho t)$$

clearly a one-parameter group. For the second,

$$\dot{\gamma}(t) = (A + B)\gamma(t)$$

gives

$$\gamma(t) = \exp((A+B)t)\gamma(0)$$
$$= I + (A+B)t + (A^2 + AB + BA + B^2)\frac{t^2}{2!} + \cdots$$

Now, $\exp(At) \cdot \exp(Bt) = I + (A+B)t + (A^2 + 2AB + B^2)t^2/2! = \alpha(t)\beta(t)$. For small enough t, then

$$\gamma(t) = \alpha(t)\beta(t),$$

is a one-parameter group. Conversely,

$$\dot{\gamma}(t) = \dot{\alpha}(t)\beta(t) + \alpha(t)\dot{\beta}(t)$$

in the neighborhood of $t = 0$, so that

$$\dot{\gamma}(t) = (A+B)\gamma(t)$$

The correspondence between multiplication in the group and addition in the algebra is visible.

To see that the commutator of matrices belongs to two vector fields and also generates a Lie group, is more difficult. Care in examining the limiting process in taking the derivative of the group element $\alpha(t)\beta(t)\alpha^{-1}9t)\beta^{-1}(t)$ gives the commutator as its infinitesimal generator, however.

It is easy to verifying these relations by examples, but first, it should be remarked that if what has been said about one-parameter groups were true only for them, then the information would be of little use. As a matter of fact, appreciation of the group properties derives from Sophus Lie's study of differential equations. The connection with useful situations is made through the action of the group on vector spaces. For example, consider a curve in $\mathbf{R}^n$ that is given by the action of a one-parameter group: $x = \alpha(t)x(0)$. Now $\dot{x}(t) = \dot{\alpha}(t)x(0) = A\alpha(t)x(0) = A_x(t)$. While Ax is a vector field on $\mathbf{R}^n$, still A is also the infinitesimal generator of a Lie group $\alpha(t)$, the determination of which corresponds to finding the solution of the differential equation. The mental picture given is of the curve $x(t)$ being traced in $\mathbf{R}^n$ by the evolution on the continuous transformation of the initial condition under the action of the group.

6.4   The Symplectic Group

Let X_A and X_B be elements of a set of vector fields $L(G)$ induced by the one-parameter subgroups $\alpha(t)$ and $\beta(t)$. $X_{\rho A}, X_{A+B}$, and X_{AB-BA} also belong to $L(G)$ and are illustrated by examples from $Sp(2)$, the group of 2×2 matrices representing the symplectic group. A matrix M belongs to $Sp(2)$ if:

$$M \begin{pmatrix} 0 & 1 \\ -1 & 0 \end{pmatrix} M^T - \begin{pmatrix} 0 & 1 \\ -1 & 0 \end{pmatrix} \tag{6.5}$$

For examples, let:

$$\alpha(t) = \begin{pmatrix} \cosh t + \sinh t & 0 \\ \sinh t & \cosh t - \sinh t \end{pmatrix}$$

$$\beta(t) = \begin{pmatrix} 1+t & t \\ -t & 1-t \end{pmatrix}$$

They both can be shown to belong to the symplectic group.

Their infinitesimal generators are:

$$A = \begin{pmatrix} 1 & 0 \\ 1 & -1 \end{pmatrix} ; B = \begin{pmatrix} 1 & 1 \\ -1 & -1 \end{pmatrix}$$

Now,

$$\rho A = \rho \begin{pmatrix} 1 & 0 \\ 1 & -1 \end{pmatrix} ; (\rho A)^2 = \rho^2 \begin{pmatrix} 1 & 0 \\ 0 & 1 \end{pmatrix}$$

Hence, exponentiating A gives:

$$\exp(\rho A t) = I \left[1 + \frac{(\rho t)^2}{2!} + \frac{(\rho t)^4}{4!} + \cdots \right] +$$

$$A \left[\rho t + \frac{(pt)^3}{3!} + \cdots \right]$$

$$= I \cosh \rho t + A \sinh \rho t = \alpha(\rho t)$$

For X_{A+B}:

$$A + B = \begin{pmatrix} 2 & 1 \\ 0 & -2 \end{pmatrix}$$

$$\exp(A + B)t = \begin{bmatrix} \sum_n (2t)^n / n! & \frac{1}{2} \sum_n (2t)^{2n+1} / (2n+1)! \\ 0 & \sum_n (-2t)^n / n! \end{bmatrix}$$

$$= \begin{bmatrix} e^{2t} & \frac{1}{2} \sinh 2t \\ 0 & e^{-2t} \end{bmatrix}$$

whose symplectic property is easily confirmed. It also agrees with $\alpha(t)\beta(t)$ to terms of first order in t. Finally,

$$AB - BA = \begin{pmatrix} -1 & 2 \\ 4 & 1 \end{pmatrix}$$

To calculate $\exp(AB - BA)t$, note that the eigenvalues of commutator are ± 3, and that the eigenvectors are $(1, 2)^T$ and $(1, -1)^T$. Hence:

$$AB - BA = \begin{pmatrix} 1 & 1 \\ 2 & -1 \end{pmatrix}^{-1} \begin{pmatrix} 3 & 0 \\ 0 & -3 \end{pmatrix} \begin{pmatrix} 1 & 1 \\ 2 & -1 \end{pmatrix}$$

and hence

$$\exp(AB - BA)t =$$
$$\begin{pmatrix} 1 & 1 \\ 2 & -1 \end{pmatrix}^{-1} \begin{pmatrix} e^{3t} & 0 \\ 0 & e^{-3t} \end{pmatrix} \begin{pmatrix} 1 & 1 \\ 2 & -1 \end{pmatrix}$$
$$= -\frac{1}{3} \begin{pmatrix} -e^{3t} - 2e^{-3t} & -e^{3t} - 1 \\ -2e^{3t} + 2e^{-3t} & -2e^{3t} - e^{-3t} \end{pmatrix}$$
$$= \begin{pmatrix} \cosh 3t - \frac{1}{3}\sinh 3t & \frac{2}{3}\sinh 3t \\ \frac{4}{3}\sinh 3t & \cosh 3t + \frac{1}{3}\sinh 3t \end{pmatrix}$$

Again, that it is symplectic and forms a one-parameter subgroup is easily verified. On the other hand, from

$$AB = \begin{pmatrix} 1 & 1 \\ 2 & 2 \end{pmatrix}$$

one gets

$$\exp(ABt) = \frac{1}{3} \begin{pmatrix} e^{3t} + 2 & e^{3t} - 1 \\ 2(e^{3t} - 1) & 2e^{3t} + 1 \end{pmatrix}$$

This proves not to be a one-parameter subgroup of $Sp(2)$ (its determinant is wrong).

If equation (6.5) is generalized from 2×2 matrices, the $2n \times 2n$ matrix M belongs to $Sp(2n, R)$ if its elements are real numbers, M^{-1} exists, and $MJM^T = J$ for

$$J = \begin{bmatrix} 0 & I_{n \times n} \\ -I_{n \times n} & 0 \end{bmatrix}$$

Then equation (40) can be written as:

$$JM^T = M^{-1}J \tag{6.6}$$

Differentiating this form of the equation (and noting that $\dot{M}M^{-1} = -M\dot{M}^{-1}$ holds because $d/dt(MM^{-1}) = 0$ holds) gives:

$$J\dot{M}^T = -M^{-1}\dot{M}M^{-1}J$$

Then $MJ\dot{M}^T = -\dot{M}M^{-1}J = JM^{-T}\dot{M}^T$ holds, by equation (40). This can be written as:

$$SJ = -JS^T \tag{6.7}$$

where S is given by $S = \dot{M}M^{-1}$, or

$$\dot{M} = SM \tag{6.8}$$

Equation (6.8) shows that S is the generator of the one-parameter symplectic group $M(t)$. Equation (6.7) says that candidates for the matrix S must have the form

$$S = \begin{bmatrix} A & BB^t \\ CC^T & -A^T \end{bmatrix}$$

that is, the off-diagonal blocks BB^T and CC^T are symmetric matrices and the diagonal blocks are related as shown.

Note that for the examples discussed earlier, with α and β symplectic, the matrices $A, B, pA, A + B$, and $AB - BA$ all have the proper S form, but AB does not.

There is an extensive theory of the relationship between Lie groups and their Lie algebras of one-parameter subgroups. This is one of those beautiful areas of mathematics where we have a category of objects (Lie groups) and a functor into another category (Lie algebras) where a great deal is known about each category. The advantage is that sometimes problems can be posed in one category, solved easily in the other, and the answer interpreted in the first. There are many such examples in mathematics but it has only been recently that such phenomena have been explicitly noted in control theory. Now, though, many such examples are evident.

Chapter 7

HOMOGENEOUS SPACES

We have at this point defined a fairly large class of objects, manifolds, groups, group actions, etc. We will now begin to consider some constructions that use several of these concepts to create new objects. We have already done this in the sense that given two manifolds we can construct a new manifold, the product manifold. Consider the old familiar concept of an equivalence relation. Recall that an equivalence relation is a relation, E, that has the property that given any object, a, in the set, S, we have aEa, given any two objects, a and b, we have a aEb iff bEa (the symmetry property), and finally the transitivity property that aEb and bEc implies that aEc. We then define S/E to be the set of all equivalence classes. That is, the set no longer consists of the individual elements that belonged to it, but only of all the classes into which those elements fall. An equivalence class of course gives a subset of S that has the property that any two elements are equivalent and that if a is an element of S and a is equivalent to an element of the equivalence class, then a is a member of the class. We now define a map $f : S \to S/E$ that takes a to $f(a)$, the equivalence class of a. It is a routine exercise to verify that f is onto. So far this is really at the level of generalized nonsense and is only suitable for a high school class in "modern mathematics". It becomes a lot more interesting when we start asking some interesting questions. For example if S is a topological space, can we define a natural topology on S/E in such a way that the map f is continuous and the space S/E inherits important properties from the topological space S? We might ask if this is true in the differentiable manifold. To the first question, we find that the answer is yes and that there is a very natural way to define the topology. To the second, we find the answer is very complex and depends on the manifold and on the equivalence relation. Let's first consider an example.

Example 7.1 *Let S be the set of all real numbers and define $E = \{(r, s) : r - s$ is an integer$\}$, i.e. two real numbers are equivalent under the relation E iff they differ by an integer. Then every equivalence class contains a real number in the interval $[0,1)$ and contains exactly one such number. On the other hand 0 and 1 are in the same equivalence class and we begin to see a picture of an interval with the end points identified. If this can be made precise, then we will have achieved the circle as an example of an S/E.*

We want the map f to be continuous and this will, in general, place restriction on the topology that can be imposed on the set S/E. For example, if we let S/E have the trivial topology with all sets being open then every equivalence class would have to be open in S. In the example of the real line this would imply that the set of integers is an open subset of the real line, which it isn't. On the other hand, if we let S/E have the topology with just two open sets then, again, every map from S to S/E is continuous and again an undesirable feature. We must reach a compromise between these two extreme topologies. *We define a set K in S/E to be open iff $f^{-1}(K)$ is open in S.* This can be verified to be a topology by *elementary arguments* using the definition of function. This, of course, is just the desired topology that makes f continuous and doesn't give too few open sets.

Going back to the first example, we identify each equivalence class with the element of $[0,1)$ which it contains. Clearly each set of the form of the open interval (r, s) is open in the set of equivalence classes. However, there are other open sets. For example the sets $[0, r)(s, 1)$ are open for every choice of real numbers r and s in the interval $[0,1)$. This is exactly the set of open sets that generates the topology on the set R/E, and we see that this is the topology of the circle. If we identify the circle with the set of complex numbers $\{\exp(2\pi i r) : r \in [0, 1)\}$ the map that takes r to $\exp(2\pi i r)$ is a topological homeomorphism. (In fact, we will see later that f is in this case a diffeomorphism.)

We will say that S/E is a *quotient space* of S and that the topology that we constructed is the *quotient topology*. The map f will be called the canonical map of S onto the quotient. We now will construct a set of examples that will serve as limits to the properties that are inherited by the quotient space.

Example 7.2 *Let S be the real plane and $E = \{(x, y) : |x| = |y|\}$. The set of equivalence classes is then just the* *set of circles centered at the origin and the origin itself. The quotient space can then be identified with the set of nonnegative real numbers, $[0, \infty)$. The topology is just the usual topology of the half line. Note that in this case the quotient is not a manifold because there cannot be a chart that contains the point 0. The quotient is, though, a nice metric space and has most of the properties needed for calculations.*

Example 7.3 *Again let S be the plane and let $E = \{(x, y) : x$ be a positive scalar multiple of $y\}$. The equivalence classes are the rays emanating from the origin and the origin itself. In this case, we can identify the quotient with the unit circle and the single point at the origin [as in the figure].*

However, the topology is different. Any open set that contains the origin must intersect each ray emanating from the origin, and hence the only open set containing the origin is the entire quotient space. Thus the quotient fails to be Hausdorf and fails to have an even more elementary separation property. In particular the quotient topology cannot be the topology induced by any metric what-so-ever.

The quotient spaces that are most often encountered in control theory are induced by group actions. Let M be a manifold and G a Lie group with a smooth action on M. In the language we have been using we let $S = M$ and define the equivalence relation $E = \{(m, n) :$ if there exists a g in G such that $gm = n\}$. Because G is a group, we can verify that E is an equivalence relation. We will often see the notation M/G for the quotient space. We see that the three examples we developed are all of this type. In Example 7.1, M is the real line and the group is the additive group of integers. In Example 7.2 the manifold is the plane and the group is the circle group. In Example 7.3 the manifold is the same and the group is the multiplicative group of positive real numbers. Thus we see that even in this case of the quotient of a manifold by a group action, the resulting topological space need not be a manifold or even a nice topological space.

Let M be a manifold and G a group acting on M with quotient space M/G. We would like to identify one unique point in each equivalence class. If we could, we would say that the set of points so identified constitute a set of *canonical forms*. If we stop and think, we see that what we have really constructed is a map from the quotient space back to the manifold M. Call this map c. We have then that a canonical form is a map c such that $f(c(\ )) = I$, the identity map on M/G. Now since M/G is a topological space we can at least ask that the map c be continuous— unfortunately this is too much to ask for in general. If we go back to Example 7.3, we find that there cannot exist such a continuous canonical form. In general we would like the map c to have as strong a set of properties as possible. For example, if M/G is a manifold we would like c to be a manifold map. Let's now consider an example of considerable more complexity, but one of real importance in control theory.

Example 7.4 *Let M be the set of all pairs of matrices (A,B) such that A is an $n \times n$ matrix, B is a $n \times n$ matrix and the system $\dot{x} = Ax + Bu$ is controllable. Let G be the group of invertible $n \times n$ matrices, $Gl(n)$. Define the group action on M in the usual manner as (PAP^{-1}, PB) so that the action is just the change of basis in the state space of the control system. The quotient space $M/Gl(n)$*

then is an object of some importance in control theory. A classical problem is to identify a set of canonical forms for this action. The canonical forms are well known and can be seen in any moderately sophisticated control theory textbook. One of the first truly new results in the applications of geometry to control theory was the study of the existence of continuous canonical forms. The proof goes somewhat as follows. First prove that the quotient space is a manifold. This is not an easy task from first principles, but there are now some easy proofs based on hard theorems in the literature. Then an invariant is attached to the quotient using results from algebraic topology and it then shows that if m differs from 1, no continuous map can exist. If m is 1, then the standard canonical form is easily seen to be continuous.

Exercise 10 *Let M be the set of square n × n matrices and consider the following group action:*

1. $A \xrightarrow{G} PAP^{-1} : P \in Gl(n)$

2. $A \xrightarrow{G} PAQ$

Discuss the quotient space and the existence of continuous canonical forms for these group actions.

An important special case of quotient topologies is the case when M is a Lie group and G is a closed subgroup. A classical but nonelementary theorem, due to the mathematician Cartan, is that the quotient space is in that case a differentiable manifold. This case is important enough that a body of notation and nomenclature has grown up around this problem. Since the rest of the world uses the notation

G/H for this quotient space we shall also. The equivalence classes are called the *left cosets* of H *in* G and the quotient space is called an *homogeneous space* of G. Example 7.1 is an example of this case with G being the real numbers and H being the subgroup of integers.

Homogeneous spaces are ubiquitous. Let M be a manifold and let G act on M. Let m be an arbitrary point in M and let C be the equivalence class of m. Then n is in C iff there exists a g in G such that $gm = n$. Let H be the set of all g in G such that $gm = m$, i.e. H is the *stabilizer of m*. Let n be an arbitrary element of C and let y be such that $ym = n$. Let h be an arbitrary element of H and consider $yhm = ym = n$. If $ym = n$ and $zm = n$, then $z^{-1}ym = z^{-1}n = m$ and therefore we have that $z^{-1}y$ is in H. Thus we can conclude that the equivalence class of m is in one to one correspondence with the set of left cosets of the stabilizer of m and therefore the orbit of m is just the space G/H. This supplies an embedding of G/H into the manifold M. Furthermore, we have that M is the union of orbits of G and hence can be written as the set theoretic disjoint union of homogeneous spaces. Of course, topologically, this can be very complex. To determine how the orbits fit together can be a very difficult problem.

The problem of determining the closure of an orbit of G has received quite a bit of attention. That is, let G/H be an orbit in M. In general the embedding is not closed and hence we can ask for the smallest closed set that contains G/H. Let's look at this problem in the examples. In Example 7.1, each orbit is of the form $\{r+n : n \text{ in } Z\}$ and hence each orbit is closed as a subset of **R**. (Group actions that have the property that the stabilizer of each element is the one point subgroup are said to be *free*.) In Example 7.2, each orbit has stabilizer equal to S^1. Thus the action is not free but every orbit is closed. In Example 7.3, every point except the origin has trivial stabilizer and the origin has stabilizer equal to the group of positive real numbers. Here the only orbit that is closed is the orbit that consists of the single point the origin. In Example 7.4, the situation is not as easy to see but a classical (and often rediscovered) fact is that each orbit is closed and the action is free. The two group actions of the problem exhibit quite different properties. The first has the property that there are infinitely many orbits and nonclosed orbits. The orbit structure of this problem is known but is really quite complicated. The second action has, of course, only finitely many orbits, and it is a rather trivial task to determine the structure of the quotient space. We leave the construction as an exercise for the reader.

Exercise 10 *Let* $\mathbf{Z}$ *denote the integers. Note that* $\mathbf{Z} \times \mathbf{Z}$ *is a subgroup of* $\mathbf{R} \times \mathbf{R}$. *Show that* $\mathbf{T}^2$ *is diffeomorphic to* $\mathbf{R} \times \mathbf{R}/\mathbf{Z} \times \mathbf{Z}$.

Chapter 8

GRASSMANNIAN TECHNIQUES

In this chapter, we develop, in some detail, a class of non-trivial manifolds, the Grassmannian manifold $G^p(V)$. The V in $G^p(V)$ is a finite dimensional vector space of dimension n. The manifold is the set of all p-dimensional subspaces of this n-dimensional space. We call it nontrivial for two reasons. One is that showing it to be a manifold is a non-trivial task that forms the heart of this chapter. The other is that these manifolds are important to control systems. It is only now being realized that they provide a foundation to many important ideas in control systems theory. This chapter should let the patient reader understand something of the manifold and believe, through seeing it applied, that it has some use.

We have seen that a C^∞ manifold is a pair (M, A) where M is a second countable Hausdorff space and A is an atlas of charts (of open sets and their maps). That $G^p(V)$ satisfies the definition of a manifold is shown in this order. First, a subspace W_A is defined, and its local coordinates and some properties described. $\Gamma(W)$ is defined as a set of W_A's. It is seen to be an open set with local maps to Euclidean space. These maps are shown to be properly diffeomorphic. Several $\Gamma(W)$'s and their maps thus form a suitable atlas for $G^p(V)$. Finally, that the $G^p(V)$ is a manifold is seen by establishing that it is Hausdorff and compact, not just 2nd countable.

A following section shows how Riccati equations arise naturally in connection with transformations on the Grassmannian. Several key properties of the equations are established though the recognition that their source is their action on these manifolds.

Before studying the Grassmannian manifold, we briefly discuss some aspects of linear optimal control. This apparent digression should help motivate the later discussion by giving some evidence that it is connected with the earlier consideration of the Lie groups and with control. It is assumed that the reader has at least an elementary knowledge of linear optimal control.

8.1   Linear Optimal Control

We discuss only the simplest problem of obtaining the optimal control of a time invariant and continuous linear dynamical system. Given the system modeled by the set of n equations

$$\dot{x} = Ax + bu \tag{8.1}$$

It is desired to choose the control $u = u^*$ so as to turn the functional:

$$J = \frac{1}{2} \int (x^T Q x + u^T R u) dt = \int \mathcal{L}(x, u) dt$$

into the smallest real number the system allows. The parameters Q and R form symmetric matrices and R is invertible.

An approach to solving this problem defines a Hamiltonian function

$$\mathcal{H}(u^*) = \min_u [\mathcal{L}(x, u) + p^T (Ax + bu)] \qquad (8.2)$$

in which an n-dimensional multiplier, p, has been introduced. The multiplier raises the dimensionality of the problem to $2n$ but allows the conditions for $(u^\times)$ to be obtained in a consistent way:

$$\partial \mathcal{H} / \partial u = 0$$

$$\dot{x}^T = \partial \mathcal{H} / \partial p$$

$$-\dot{p}^T = \partial \mathcal{H} / \partial x \qquad (8.3)$$

Applying the recipe of equations (8.3) to (8.2) and transposing terms give the equations

$$u = -R^{-1} b^T p$$

$$\dot{x} = Ax + bu = Ax - b R^{-1} b^T p$$

$$-\dot{p} = Qx + A^T p \qquad (8.4)$$

The last two equations in (8.4) are called Hamilton's equations. They can be written in matrix form as:

$$\frac{d}{dt}\begin{bmatrix} x \\ p \end{bmatrix} = \begin{bmatrix} A & -bR^{-1}b^T \\ -Q & -A^T \end{bmatrix}\begin{bmatrix} x \\ p \end{bmatrix} = H\begin{bmatrix} x \\ p \end{bmatrix} \qquad (8.5)$$

The matrix H in equation (8.5) is sometimes called the Hamiltonian of optimal control. It has the right form to be the generator of a one-parameter symplectic group. The equation is solved for initial conditions on $x(t)$ and final conditions on $p(t)$.

The complications due to solving these coupled equations with mixed boundary conditions can be made someone else's problem by a transformation:

$$\begin{bmatrix} \xi \\ \eta \end{bmatrix} = \begin{bmatrix} I & 0 \\ -P & I \end{bmatrix}\begin{bmatrix} x \\ p \end{bmatrix} ; \begin{bmatrix} x \\ p \end{bmatrix} = \begin{bmatrix} I & 0 \\ P & I \end{bmatrix}\begin{bmatrix} \xi \\ \eta \end{bmatrix} \qquad (8.6)$$

The matrices I and P in equation (8.6) are $n \times n$. If P is symmetric, the character of the Hamiltonian matrix as the symplectic generator will be preserved.

Applying the transformation of equation (8.6) to (8.5) gives:

$$\frac{d}{dt}\begin{bmatrix} \xi \\ \eta \end{bmatrix} = \qquad (8.7)$$

$$\left\{ \begin{bmatrix} 0 & 0 \\ -\dot{P} & 0 \end{bmatrix} + \begin{bmatrix} I & 0 \\ -P & I \end{bmatrix}\begin{bmatrix} A & -bR^{-1}b^T \\ -Q & -A^T \end{bmatrix} \right\} \times$$

$$\begin{bmatrix} I & 0 \\ P & I \end{bmatrix}\begin{bmatrix} \xi \\ \eta \end{bmatrix}$$

$$= \begin{bmatrix} A - bR^{-1}b^TP & -bR^{-1}b^T \\ Z(P) & -(A - bR^{-1}b^TP)^T \end{bmatrix} \begin{bmatrix} \xi \\ \eta \end{bmatrix} \quad (8.8)$$

where $Z(P) = -\dot{P} - PA - A^TP + PbR^1b^TP - Q$.

Examining equation (50) shows that it leads to the decoupled equation $\dot{\xi} = (A - bR^{-1}b^TP)\xi$ by choosing P so that $\eta = 0$ at least at one time and so that $\dot{\eta} = 0$ holds. These choices imply the conditions:

$$-\dot{P} = PA + A^TP - PbR_{-1}b^TP + Q \quad (8.9)$$

and

$$p(t_0) = P(t_0)x(t_0) \quad (8.10)$$

Equation (8.9) is a matrix Riccati equation. Equation (8.10) shows that the multiplier constrains $x(t)$ to a curve on an n-dimensional manifold. Though the manifold of interest is the one which supports the trajectory of the controlled dynamical system, it is characterized by the selection of p as a complement to x in the $2n$-dimensional space.

This procedure may seem to be a novel way of selecting the multiplier to solve the optimization problem. In fact, it is an illustration of an approach to the study of geometry which was introduced by Grassmann in the 1840's. Following the earlier studies of projective space, he looked into the relations between a curve given in one space by a set of parameters and the curve it induced in the parameter space. In our example, $x(t)$ is a curve with certain parameters which describe a curve in a dual space. The transformation is generated by the Riccati equation. In a general Grassmannian, the two spaces need not have the same dimension.

8.2   The Grassmannian

Showing the Grassmannian $G^p(V)$ to be a manifold is developed in this section. Now, $G^p(V)$ is the set of all p-dimensional subspaces of an n-dimensional vector space, V. It is covered by an atlas of charts $\Gamma(W)$. Each $\Gamma(W)$ is the set of subspaces W_A given by the local coordinate map written as the $(n-p) \times p$ matrix, A. First, a simple example will be given to fix ideas of the space V and subspace W_A. Then the nomenclature is explained in detail.

A simple example

Take V as the plane $\mathbf{R}^2$, and let U and W be one dimensional subspaces of it such that V is their Cartesian product: $V = U \times W$. Now, U and W are to have no subspaces in common, but since they are to be vector spaces, they do share the point (0,0) of V in this representation, which will be referred to as the set $\{0\}$. The situation can be visualized as in sketch (h), where one can think of the plane V with oblique coordinate directions U and W defined in the sketch.

The point P in the sketch belongs to the subspace W_A. It is specified uniquely by $u + Au$, with A a real number (a 1×1 matrix) and $u \in U$. Every point in W_A is specified similarly by some u, and every point in the plane except W itself belongs to a W_A for some A. The collection of these W_A for the given choice of W is an open set $\Gamma(W)$. W is included in a chart for a different subspace $\overline{W}$ of V, where now $V = U \times \overline{W}$. Then W is, say, the subspace $\overline{W}_{\bar{A}}$ and belongs to the open set $\Gamma(\overline{W})$. Now W_A and $\Gamma(W)$ will be described in general.

The subspace W_A

Let U and W be subspaces of V such that the dimension of U is p, the dimension of W is $n - p$, and V is their direct sum. By the direct sum is meant that U and W have no subspace in common; their intersection contains no subspace. Consider the subspaces W_A such that:

$$W_A = [u + Au : u \in U \text{ and } A \in L(U, W)]$$

Now note that the dimension of W_A is p; let $u_1, \cdots, u_p$ be a basis for U, and suppose $\alpha_1(u_1 + Au_1) + \cdots + \alpha_p(u_p + Au_p) = 0$. Then

$$\alpha_i u_i = -\sum \alpha_i A u_i$$

and

$$\sum \alpha_i u_i = 0$$

since $V = U \oplus W$. Thus, all of the $\alpha_i = 0$ and we conclude that $\{u_i + Au_i : i \leq p\}$ is linearly independent and is a basis for W_A. The space W_A is a p-dimensional subspace determined by the linear transformation A.

The decomposition of W_A is unique because U and W have zero intersection; suppose $W_A = W_B$, then $u_i + Au_i = v_i + Bv_i$ for some choice of v_i and we can assume that the u_i form a basis for U. Then $u_i - v_i = Bv_i - Au_i$. This implies that $u_i = v_i$ because $u_i - v_i \in U, Bv_i - Au_i \in W$, and $U \cap W = 0$. We conclude that $Au_i = Bu_i$ for a basis of U; thus $A = B$.

Not every p-dimensional subspace of V can be represented as a W_A. We can prove the following fact which will be useful later:

A p-dimensional subspace of V can be represented as a W_A for some A iff $U_o \cap W = \{0\}$.

Proof: Let $z \in U_o \cap W$; if $U_o = W_A$ then $z = u + Au$ for some u and so $u = z - Au$. Since $z \in W$ and $Au \in W$ we have that $u \in W$, hence, is zero. Thus, $U_o = W_A$ implies that $U_o \cap W = \{0\}$. On the other hand, suppose $U_o \cap W = \{0\}$. Let $s_a, \cdots, s_p$ be a basis for U_o and write each s_i as $s_i = u_i + w_i$. This decomposition is unique. Now, the u_i form a basis for U; for, suppose that

$$\sum \alpha_i u_i = 0$$

Then we have that

$$\sum \alpha_i s_i = \sum \alpha_i w_i$$

which implies that

$$\sum \alpha_i s_i = 0$$

so that $\alpha_i = 0$ for all i. Thus, $\{u_i : 1 \leq p\}$ is a basis for U, and no subspaces are left in U_o to be generated by any of the w_i. Let A be the linear map that takes u_i to w_i. Then $U_o = W_A$.

The charts $\Gamma(W)$

Let $\Gamma(W)$ be the set of all W_A. Thus $\Gamma(W)$ is indexed by the set of linear maps, $L(U, W)$ from U into W and we've

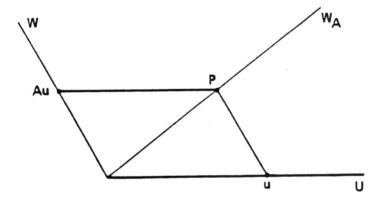

Figure 8.1: Sketch(h)

seen that each element of $\Gamma(W)$ is uniquely associated with an element of $L(U,W)$. We can define a map from $\Gamma(W)$ to $\mathbf{R}^{(n-p)\times p}$ by

$$\phi(W_A) = A.$$

The map ϕ depends on a choice of basis for the space V. It can be shown that every p-dimensional subspace of V is in $\Gamma(W)$ for some choice of W. Then $(\Gamma(W), \phi)$ are suitable candidates for chart mappings. Their differentiable structure needs to be verified.

With reference to sketch (i), let S be an element of $\Gamma(W)$ and $\Gamma(W')$ and suppose $S = W_A$. We need to determine a T such that $S = W'_T$. Each element w of W can be written uniquely as a sum of elements from W' and U. As illustrated in sketch (j), let $w = u_i + w'_o$. Define two functions A_1 and A_2 by $A_1(w) = u_1$ and $A_2(w) = w'_1$. Since the decomposition is unique, it follows that $A_1 \in L(W,U)$ and $A_2 \in L(W,W')$. It will be useful later to note now that A_2 is invertible. This is shown by supposing $A_2(W) = 0$. Then $w = A_1(w)$. But this implies that $w \in U$, which is equivalent to $w = 0$. Thus, A_2 is one to one, and since W and W' have the same dimension, A_2 is invertible. Recall that:

$$S = W_A = \{u + Au : A \in L(U,W)\}. \tag{8.11}$$

From what has just been derived, we can write:

$$\left\{ \begin{array}{rl} W_A = & \{u + A_1Au + A_2Au : A \in L(U,W)\} \\[2mm] = & \{(I + A_1A)u + A_2Au : A \in L(U,W)\} \\[2mm] = & \{u' + A_2A(I + A_1A)^{-1}u' : A \in L(U,W)\} \end{array} \right\} \tag{8.12}$$

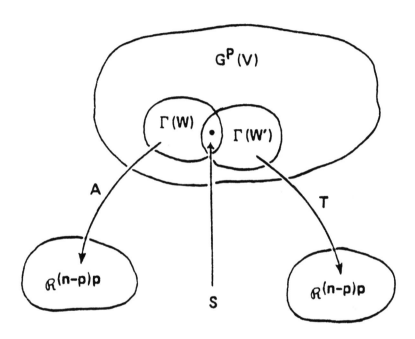

Figure 8.2: Sketch (i)

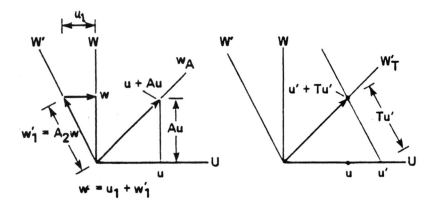

Figure 8.3: Sketch (j)

Since $A_2 A(I + A_1 A)^{-1} \in L(U, W')$, we have that

$$W_A = W'_{A_2 A(I+A_1 A)-1} \tag{8.13}$$

Since the derivation can be run backwards, it can be seen that if $S \in \Gamma(W) \cap \Gamma(W')$, then the inverse actually exists. Since the coordinate change is a rational function with nonzero denominator, it is C^∞. The significance of its form will become clear later.

The manifold $G^p(V)$

The sets Γ have now been defined. It was said that every p-dimensional subspace belongs to some $\Gamma(W)$, because some appropriate complementary subspace W can be found for it. Therefore, an appropriate map is obtained. Hence $G^p(V)$ is covered by the sets $\Gamma(W)$. It remains to show that $G^p(V)$ is Hausdorff and compact.

It is easy to see that it is Hausdorff. This follows from the fact that given two p-dimensional subspaces, S_1 and S_2, there exists a W such that S_1 and S_2 are both in $\Gamma(W)$. Therefore, both are in the same $\mathbf{R}^{(n-p) \times p}$, which is Hausdorff. Thus $G^p(V)$ is indeed an $(n - p) \times p$-dimensional manifold.

The proof that $G^p(V)$ is compact relies on the theorem that the image of a continuous map of a compact set is compact. Now, the unitary group, $U(n)$, is compact. The strategy is to show that there is a continuous map, T, from $U(n)$ to $G^p(V)$ so that $G^p(V)$ inherits a natural structure from $U(n)$. If the inverse image of any open set $\mathcal{O} \cap \Gamma(W)$ of $G^p(V), T^{-1}(\mathcal{O})$, is open in $U(n)$, then T is continuous and compactness will follow.

Define T from $U(n)$ to $G^p(V)$ by:

$$T(\alpha) = \alpha(U) \tag{8.14}$$

where U is the p-dimensional subspace of the definition of $G^p(V)$. Since $\alpha \in U(n), \alpha(U)$ is p-dimensional and so is in $G^p(V)$. It remains to show that T is continuous.

Let $\mathcal{O}$ be an open subset of some chart $\Gamma(W)$. To prove continuity, it suffices to show that $T^{-1}(\mathcal{O})$ is open. Since $\mathcal{O}$ is contained in $\Gamma(W), \mathcal{O}$ is of the form $\{W_A : A \in \mathcal{O} \cap \mathbf{R}^{(n-p)p}, \mathcal{O}$ open $\}$. Now,

$$T^{-1}(W_A) = T^{-1}\{(x, Ax) : x \in U\}$$

so we look for the α such that $\alpha\{(x,0) : x \in U\} = \{(x, Ax) : x \in U\}$. Partitioning α as

$$\left[\begin{array}{c|c} \alpha_1 & \alpha_2 \\ \hline \alpha_3 & \alpha_4 \end{array} \right] \tag{8.15}$$

Where α is $p \times p, \alpha_2$ is $p \times (n-p), \alpha_3$ is $(n-p) \times p$ and α_4 is $(n-p) \times (n-p)$, we have

$$\begin{aligned} \alpha(x) &= \{(\alpha_1 x, \alpha_3 x) : x \in U\} \\ &= \{(x, \alpha_3 \alpha_1^{-1} x) : x \in U\} \tag{8.16} \end{aligned}$$

$$T^{-1}(\mathcal{O}) = \{\alpha : \alpha_3 \alpha_1^{-1} \in \mathcal{O}'\} \tag{8.17}$$

Continuity of matrix multiplication assures us that the set of *all* matrices

$$\{(\alpha_1, \alpha_3) : \alpha_3 \alpha_1^{-1} \in \mathcal{O}'\} \qquad (8.18)$$

is open as a subset of $\mathbf{R}^{n(n-p)}$; hence, the set of *all* matrices

$$\begin{bmatrix} \alpha_1 & \alpha_2 \\ \alpha_3 & \alpha_4 \end{bmatrix} \qquad (8.19)$$

with $\alpha_3 \alpha_1^{-1} \in \mathcal{O}'$ is open as a subset of $\mathbf{R}^{n^2}$. Thus, the intersection of this open set with $U(n)$ is open and $T^{-1}(\mathcal{O})$ is an open set. So we have that T is a continuous map from $U(n)$ onto $G^p(V)$. Since $U(n)$ is compact and T is continuous, $G^p(V)$ is compact.

We have now constructed a manifold from a rather complicated set. It figures prominently in several sets of developments in control theory, one of which concerns solutions of the Riccati equation.

8.3 An Application

As an application of the Grassmannian which is of some importance to those reading the control theory literature, this section discusses how the Riccati equation arises naturally in this context and what some of its properties are.

The Riccati equation

It was seen earlier in this section that the Riccati equation arose in connection with the transformation of Hamilton's equation. Here we will see that it arises quite generally as the generator of the transformation of the Grassmann manifold. First, the general transformation will be given. Then its dynamic behavior will be seen to be governed by the Riccati equation.

To find the transformation, we examine the action of the general linear group, $Gl(n)$, on $G^p(V)$. Let $\alpha \in Gl(n)$ and partition α as in (8.15). Note that an action of $Gl(n)$ on $G^p(V)$ is well defined by:

$$(\alpha, W) \rightarrow \alpha(W)$$

where W is a p-dimensional subspace of V. The local representative of this action is of interest.

$$\alpha(W_A) = \alpha(\{(x, Ax) : x \in U\})$$

$$= \{(\alpha_1 x + \alpha_2 Ax, \alpha_3 x + \alpha_4 Ax) : x \in U\}$$

Now, if $\alpha(W_A) \in \Gamma(W)$, then there is a B such that

$$\alpha(W_A) = \{(z, Bz) : z \in U\}.$$

In particular, $(\alpha_1 + \alpha_2 A)x = z$ has a solution for all $z \in U$ and hence $(\alpha_1 + \alpha_2 A)^{-1}$ exists. Thus

$$\alpha(W_A) = \{(z, (\alpha_3 + \alpha_4 A)(\alpha_1 + \alpha_2 A)^{-1}z) : z \in U\}$$

$$= W_{(\alpha_3 + \alpha_4 A)(\alpha_1 + \alpha_2 A)^{-1}} \tag{8.20}$$

whenever $(\alpha_1 + \alpha_2 A)^{-1}$ exists and the inverse exits if and only if $\alpha(W_A) \in \Gamma(W)$. Also note that the inverse exists iff $\alpha(W_A)$ has zero intersection with W.

The function

$$\alpha : A \rightarrow (\alpha_3 + \alpha_4 A)(\alpha_1 + \alpha_2 A)^{-1} \tag{8.21}$$

is called a *generalized linear fractional transformation*, and is the transformation in parameter space that was sought.

To examine its dynamic behavior, let $t \rightarrow \alpha(t)$ be a one-parameter subgroup of $Gl(n)$, and let

$$B(t) = \left(\frac{d\alpha}{dt}(t) \right) \alpha(t)^{-1} \qquad (8.22)$$

be its infinitesimal generator. Let W_A be some fixed element of $G^p(V)$. We have that $t \rightarrow \alpha(t)W_A$ is a curve in $G^p(V)$. Let us calculate the vector field with which it is associated. Writing α as before, we have $\dot{\alpha} = B\alpha(t)$:

$$\begin{bmatrix} \dot{\alpha}_1 & \dot{\alpha}_2 \\ \dot{\alpha}_3 & \dot{\alpha}_4 \end{bmatrix} = \begin{bmatrix} B_{11}\alpha_1 + B_{12}\alpha_3 & B_{11}\alpha_2 + B_{12}\alpha_4 \\ B_{21}\alpha_1 + B_{22}\alpha_3 & B_{21}\alpha_2 + B_{22}\alpha_4 \end{bmatrix} \qquad (8.23)$$

From (8.21) we see that we must calculate

$$\frac{d}{dt}[(\alpha_3 + \alpha_4 A)(\alpha_1 + \alpha_2 A)^{-1}] \qquad (8.24)$$

Using the recipe:

$$\frac{d}{dt}XY^{-1} = \dot{X}Y^{-1} - XY^{-1}\dot{Y}Y^{-1}$$

gives for (8.24):

$$(\dot{\alpha}_3 + \dot{\alpha}_4 A)(\alpha_1 + \alpha_2 A)^{-1}$$

$$- (\alpha_3 + \alpha_4 A)(\alpha_1 + \alpha_2 A)^{-1}(\dot{\alpha}_1 + \dot{\alpha}_2 A)(\alpha_1 + \alpha_2 A)^{-1} \quad (8.25)$$

Substituting for the $\dot\alpha_i$ from (8.23) and collecting terms in (8.25) gives:

$$[B_{21}\alpha_1 + B_{22}\alpha_3 + (B_{21}\alpha_2 + B_{22}\alpha_4)A](\alpha_1 + \alpha_2 A)^{-1}$$
$$-(\alpha_3 + \alpha_4 A)(\alpha_1 + \alpha_2 A)^{-1}[B_{11}\alpha_1 + B_{12}\alpha_3+$$
$$(B_{11}\alpha_2 + B_{12}\alpha_4)A](\alpha_1 + \alpha_2 A)^{-2} =$$
$$B_{21} + B_{22}(\alpha_3 + \alpha_4 A)(\alpha_1 + \alpha_2 A)^{-1} - (\alpha_3 + \alpha_4 A)\times$$
$$(\alpha_1 + \alpha_2 A)^{-1}[B_{11} + B_{12}(\alpha_3 + \alpha_4 A)(\alpha_1 + \alpha_2 A)^{-1}] =$$
$$B_{21} + B_{22}(\alpha W_A) - (\alpha W_A)B_{11} - (\alpha W_A)B_{12}(\alpha W_A)$$

Thus, we show that the curve $\alpha(t)W_A$ satisfies the differential equation:

$$\dot P = B_{21} + B_{22}P - PB_{11} + PB_{12}P \qquad (8.26)$$

$$P(0) = A$$

Thus, Riccati differential equations are associated with a natural group action on the manifold $G^p(V)$.

Some properties of Riccati equations

There are properties of the Riccati equations that can be deduced just from knowledge of the manifold and the fact that they are associated with a group action. For example, the Grassmannian manifold $G^p(V)$ shares with the sphere the property that every vector field vanishes at least at one point; i.e., there is a point $W \in G^p(V)$ such that $A(t)W = W$ for all t.

Now if $W \in \Gamma(W')$ then with respect to the local coordinates defined by $\Gamma(W')$ we have that

$$0 = \dot{P}(t) = B_{21} = B_{22}P - PB_{11} - PB_{12}P \qquad (8.27)$$

and hence, there is a solution of the algebraic Riccati equation. However, given the algebraic Riccati equation (8.27) there may *not* be a matrix P that satisfies (8.27). The guaranteed solution may not be in the required chart, i.e., the solution exists at '∞'.

In the rest of this section we will study some of the elementary consequences of the fact that Riccati differential equations are associated with a group action on $G^p(V)$. As a first problem we ask:

Problem 1. What are necessary and sufficient conditions for the algebraic Riccati equation to have a solution?

By the algebraic Riccati equation, we mean the equation

$$B_{21} + B_{22}P - PB_{11} - PB_{12}P = 0 \qquad (8.28)$$

An equivalent question is: what are the constant solutions of (8.27)?

The general theory tells us that every vector field on $G^p(V)$ vanishes at least at one point so we know that in the large there are solutions although they may be 'solutions at infinity'. So instead of searching for solutions of (8.28), we can look for p-dimensional subspaces, W, of V such that

$$\alpha(t)W_A = W_A \qquad (8.29)$$

for all $t \in \mathbf{R}$. Unfortunately, all we know about $\alpha(t)$ is that it has infinitesimal generator B. Thus, we need the following lemma.

Lemma 8.1 *Let $\dot{\alpha}(t) = B\alpha(t)$ and let $W \in G^p(V)$. Then $\alpha(t)W = W$ for all $t \in \mathbf{R}$ if $BW \subseteq W$.*

Proof: Suppose $\alpha(t)W = W$. Let $w_i, i = 1, \cdots, p$ be a basis for W, then $\alpha(t)w_i = \sum_{j=1}^p \alpha_{ij}(t)w_j$ and we have as a consequence that $\alpha(t)w_i = \sum \dot{\alpha}ij(t)w_j$ so that $\dot{\alpha}(t)W \subset W$. Thus,

$$B\alpha(t) = BW$$

and

$$B\alpha(t)W = \dot{\alpha}(t)W \subseteq W$$

so that $BW \subseteq W$. On the other hand, assume that $BW \subseteq W$. Now $\alpha(t) = \exp(Bt)$. Using the definition of $\exp(Bt)$ and the fact that $B^kW \subseteq W$ for all k it follows that $\alpha(t)W = W$.

The lemma reduces the problem of finding invariant subspaces of $\alpha(t)$ to finding invariant subspaces of B, technically at least, an easier problem. We can now restrict ourselves to the study of the invariant subspaces of B. Our problem has been reduced to asking if there is a p-dimensional invariant subspace of B that has zero intersection with U_0 (recall chapter 8.2.2). We answer this question first in the generic case.

Assume that B has n linearly independent eigenvectors $\eta_1 \cdots, \eta_n$. We claim that in this case there is always such a solution. Assume $\eta_1 \notin U_o$. There is such an η_1 for if not the n eigenvectors span a p-dimensional space that contradicts their independence. Assume V_k has been constructed such that $V_k = < \eta_1, \cdots, \eta_k >$ and $V_k \cap U_o = \{0\}$. If $k = p$ we are finished, if not, then $k < p$ (k cannot be greater than p and $V_k \cap U_o = \{0\}$). Let $W_r = < V_k, \eta_r >$, for $r > k$ and suppose

$W_r \cap U_o \neq \{0\}$ for all $r > k$. Then we have that

$$\sum_{i=1}^{k} \alpha_{ij}\eta_i + \alpha_j\eta_j = u_j, \; j = k+1, \cdots, n \qquad (8.30)$$

There are more than $n - p$ equations and thus there are nonzero a_i such that

$$\sum_{j=k+1}^{n} a_j \left(\sum \alpha_{ij}\eta_j + \alpha_j\eta_j\right) = \sum a_j u_j = 0$$

therefore, $a_j\alpha_j = 0$ for $j = k+1, \cdots, n$ because of independence. We have that $a_j = 0$ since α_j cannot be zero by hypothesis but this contradicts the existence of such a set of u_j. So for some r

$$W_r \cap U_0 = \{0\}$$

Renumbering if necessary, let $V_{k+1} = W_r$. We eventually have a V_k such that $V_k \oplus U_0 = V$; hence, there is a P such that $V_k = W_p$ and P is the desired solution of (8.28), Potter, [13], essentially recognized this fact.

A second problem can be considered. We have long associated various groups with differential equations and have asked what group leaves some properties or form of the solution of a differential equation invariant? Of course, in that imprecise form there is no precise answer but maybe an example will clarify the issue somewhat. Consider an ordinary linear differential equation

$$\dot{x} = Ax \qquad (8.31)$$

We are all used to the concept of changing basis in state space to transform the equation to

$$\dot{z} = (\alpha A \alpha^{-1})z \qquad (8.32)$$

Such a transformation has the useful property that it leaves the linearity invariant. Let us now examine the following problem.

Problem 2. Is there a large group of transformations that leaves the quadratic nature of the Riccati differential equation invariant.

Now this problem is perhaps a little harder than it would first appear. Consider again equation (8.27) and make the transformation

$$P = \alpha S$$

where $\alpha \in Gl(n - p)$. Then S satisfies

$$\dot{S} = (\alpha^{-1}B_{21}) + (\alpha^{-1}B_{22}\alpha)S - S(B_{11}) - S(B_{12}\alpha)S \quad (8.33)$$

and so the Riccati nature is preserved. However, it is well known in the control-theory literature that if

$$S = P^{-1}$$

then S again satisfies a Riccati equation, for

$$PS = I$$

implies

$$\dot{P}S + P\dot{S} = 0$$

which yields

$$P^{-1}\dot{P}S + \dot{S} = 0$$

and

$$\dot{S} = B_{12} + B_{11}S - SB_{22} - SB_{21}S \quad (8.34)$$

Thus, the group we are seeking is larger than just a linear change of basis. In fact we will demonstrate the following:

Theorem 8.1 *The class of Riccati equations is invariant under transformation by generalized linear fractional transformations: i.e., if*

$$S = (\alpha_1 P + \alpha_2)(\alpha_3 P + \alpha_4)^{-1}$$

then, if P satisfies a Riccati equation, so does S.

To prove this, we will work in the global situation rather than in the local coordinates. Let $P(t)$ be the solution of (67), and let $A(t)$ be the one-parameter subgroup such that

$$W_p(t) = A(t)W_X \qquad (8.35)$$

Let B be the infinitesimal generator of $A(t)$. From (8.21) we know that linear fractional transformations of P are associated with the action of $Gl(n)$ on $G^p(V)$ and so we are asking what (if any) differential equation does the curve

$$t \to \alpha W_{p(t)} \qquad (8.36)$$

satisfy?

Now obviously

$$\alpha W_{p(t)} - \alpha A(t)W_X$$

but $\alpha A(t)$ is *not* a one-parameter subgroup (it is not a subgroup). However, we do have

$$\alpha W_{p(t)} = \alpha A(t)\alpha^{-1}(\alpha W_X) \qquad (8.37)$$

and $\alpha A(t)\alpha^{-1}$ is a subgroup. This is true if

$$\alpha = \begin{bmatrix} \alpha_{11} & \alpha_{12} \\ \alpha_{21} & \alpha_{22} \end{bmatrix}$$

and

$$S(t) = (\alpha_{21} + \alpha_{22}P(t))(\alpha_{11} + \alpha_{12}P(t))^{-1}$$

then

$$W_{S(t)} = \alpha A(t)\alpha^{-1}(\alpha W_X)$$

and so $S(t)$ satisfies a Riccati differential equation. The infinitesimal generator of

$$A(t)\alpha^{-1}$$

is $\alpha B\alpha^{-1}$ and so the coefficients of the Riccati equation of S are related to the coefficients of the Riccati equation of P by

$$B \rightarrow \alpha B\alpha^{-1}$$

This *not* easy to show directly.

Following this line of thought, one can mention two special cases. In the first case, suppose B has n distinct eigenvectors. Then, there is an α such that $\alpha B\alpha^{-1}$ is diagonal and the associated Riccati equation is

$$\dot{S} = -D_1 S + SD_2$$

where

$$\alpha B\alpha^{-1} = \begin{bmatrix} D_1 & 0 \\ 0 & D_2 \end{bmatrix}$$

Thus, B is equivalent to the study of simpler linear differential equations.

For the second case, consider the fact that every B is equivalent to a matrix of the form

$$\begin{bmatrix} B_1 & 0 \\ E & B_2 \end{bmatrix}$$

where E is the $(n-p) \times p$ matrix, which is zero except possibly for the element at the 1, p position of E. Thus, every Riccati equation can be transformed into a linear equation

$$\dot{S} = E + SB_2 - B_1S$$

This fact is rather elementary but doesn't appear to have been noted in the literature.

There are many results that can be proven about Riccati equations using the Grassmannian techniques. However, they belong more properly to a research monograph than to an introduction to differential geometry.

Chapter 9

CONCLUDING REMARKS

We have attempted in this report to give an informal introduction to differential geometry that would be palatable for a nonmathematically trained engineer. It is primarily intended for the control engineers, but we hope that persons in other disciplines are interested in learning these basic facts.

Anyone who has read these notes by now realizes that a lot has been left unsaid. If one tries to apply differential geometric techniques to real problems he or she will quickly see that this material must be augmented by more powerful results. The purpose of these notes is to prepare the reader to delve into the more specific areas of differential geometry.

Many excellent books are available. The authors have found the following to be particularly useful. *Calculus on Manifolds* by Spivak, [4], should be read by anyone interested in differential geometry. It is short and concise, setting the stage for serious study.

An Introduction to Differentiable Manifolds and Riemannian Geometry by W. Boothby, [2], is readable and is a first class source book, as is *Differential Geometry and the Calculus of Variations* by R. Hermann, [1]. There are many other introductory books on differential geometry available — some are quite good and others are only mediocre. The problem with most is that they were written for persons assumed to have good mathematical background as well mathematical sophistication.

At the more specialized level, the monographs on mechanics by Abraham and Marsden, [14], and the monograph by Arnold, [15] are available. Both are advanced and sophisticated but are very well written. Both books deserve to be in everyone's reference library.

For control theory, one should be familiar with the various books of R.Hermann. In addition, one must consult the current literature. The SIAM Journal of Control and the IEEE Transactions on Automatic Control both contain occasional papers on geometric control theory. Many of the papers at a Harvard workshop sponsored by NASA-Ames, NATO, and the AMS, [16, 17], have the concepts of manifolds and of differential geometry at their core.

One should remember that control theory problems do not arise for the benefit of differential geometry and *every* available method should be used to obtain a satisfactory solution. Differential geometry is just a tool and the control engineer should not restrict himself to one tool but should be familiar with as many as possible.

Chapter 10

APPENDIX: VECTOR CALCULUS

This appendix recalls a number of concepts usually covered in an advanced calculus course. Though it might serve to prepare the reader psychologically, its true purpose is simply to express explicitly the ground rules for the contents of the body of the book.

10.1   Real Euclidean Space

The real linear vector n-space, $\mathbf{R^n}$, is the set of all n-tuples of real numbers. That is,

$$\mathbf{R}^n = \{(x, \cdots, x_n) : x_i \in \mathbf{R}, i = 1, \cdots, n\}$$

These numbers, when added together such that $x + y = (x_1, +y_1, \cdots, x_n + y_n)$ or multiplied by real numbers so $ax =$

$(ax_1, \cdots, ax_n)$ for $a \in \mathbf{R}$ give results that also belong to the same space $\mathbf{R^n}$. A complex vector space is defined accordingly, and is denoted by $\mathbf{C}^n$.

The vector space becomes a real euclidean space (also often denoted by $\mathbf{R^n}$) when it has a particular measure of size, a norm. This norm is just a generalization of the concept of length in ordinary 3-space, $\mathbf{R^3}$. For $x \in \mathbf{R}^n$, we define the *norm* of x as the real number:

$$|x| = (x_1^2 + \cdots + x_n^2)^{1/2}$$

Intuition holds, and $|()|$ has all the properties of length that one expects.

The *inner product* of two vectors in $\mathbf{R}^n$, x and y, can be defined as:

$$x \cdot y = x_1 y_1 + \cdots + x_n y_n$$

The norm and inner product functions are related by:

$$x \cdot x = |x|^2$$

We will say that x is *orthogonal* to y if

$$x \cdot y = 0$$

10.2   Topological Spaces

Even though the concepts of norm and inner product are essentially vector space concepts, they give rise to a convenient mode of expression of topological properties that are necessary to the development of calculus.

An *open ball* of radius r at $x_0, S_r(x_0)$, is the set of all those points lying within a distance r of the point x_0:

$$S_r(x_0) = \{x : |x - x_0| < r\}$$

A subset U of $\mathbf{R}^n, U \subset \mathbf{R}^n$, is said to be *open* if, for each x in U there is an r such that $S_r(x) \subset U$; that is, U contains an open ball about each of its points. A subset C of $\mathbf{R}^n$ is said to be *closed* if its complement, that is, the set of all x that do not belong to $C, \{x : x \notin C\}$, is an open subset. The collection of all such open subsets of $\mathbf{R}^n$ constitutes one possible *topology* for $\mathbf{R}^n$. We refer to it as the usual topology, the Euclidean topology, or as the metric topology, depending on how precise we feel like being at the moment. The discussion usually concerns $\mathbf{R}^n$, perhaps with the Euclidean notion of distance. There are occasions later in this section when we refer to objects that are topological spaces but which are not necessarily metric subspaces of $\mathbf{R}^n$.

A typical exercise throughout this paper is to refer whatever object that is being studied back to the real numbers or to spaces constructed of strings of real numbers. The reason is simply that we know the properties of real numbers. If we can relate the object under study to the real numbers by a continuous function with a continuous inverse, then we can transfer fundamental properties of the real numbers to the object of interest.

What the fundamental properties are that are needed for calculations and how objects differ by having different ones of these properties is the concern of point set topology. Our interest in topology is simply to say what are some requisites for the operations of differential geometry. Since the treatment is introductory, we do not discuss the various classes of geometric objects that one can come upon.

Our use of topology, then, is like the scientist's or engineer's interest in fundamental physical standards like the standard meter or the standard kilogram. To make a comparison of length, the scientist lays the standard along his rod which he marks according to the standard's marks. The mathematical equivalent of this laying alongside, is finding a continuous function that relates an open set of the real numbers and an open set of the other mathematical object. If one matching satisfies for the whole object, fine. Usually the comparison is done in pieces to prevent uncertainty and equivocal results using as many open sets as are needed. The type of number of open sets required is important mathematically and helps classify the geometric object. We need a finite or, at worst, a countable number of them.

A *topological space*, in general, is a pair (X, Ω) where X is a set and Ω is a collection of subsets of X with the property that the empty set, ϕ, is in Ω, the intersection of any two elements of Ω is in Ω, and the union of an arbitrary number of members of Ω is also in Ω. The elements of Ω are called the open sets for the topological space (X, Ω). For example, $\mathbf{R}^n$ with an Ω consisting of sets that are open in the sense defined in the previous paragraph, $(\mathbf{R}^n, \Omega)$, is a topological space (with the usual topology).

Two extreme examples of topological spaces can be constructed out of *any* set X; namely, when $\Omega = \{\phi, X\}$ and when $\Omega = \{\text{all subsets of } X\}$. In the first, the 'indiscrete' case, Ω, containing only the empty set ϕ, has too few sets to be very useful; in the second, it has too many. Requiring that the following two conditions hold for the space eliminates those two extremes from further consideration.

The first condition to be imposed is that the space (X, Ω) be Hausdorff: given any two distinct points x and y in X, there are two open sets U, and V such that $x \in U, y \in V$, but their intersection is empty, $U \cap V = \phi$. This condition is essential for analysis, because it assures us that sequences converge to unique limits. Since it also guarantees that Ω has a lot of open sets, it rules out the indiscrete topology from selection.

The second condition to be imposed is that the space (X, Ω) be second countable; that is, that there be a countable subset of Ω, $S = \{S_1, S_2, \cdots\}$ such that every element of Ω is the union of a finite number of intersections of elements of S. Then X is said to have a countable basis of open sets. Thus, this condition assures us that there aren't too many open sets. $\mathbf{R^n}$ is an example of a second countable topological space and the metric topology. A suitable set S for this Ω is the set of all open balls $S_r(x)$ where r and the coordinates of x are rational numbers. The set S is a countable set since it is a countable set of countable sets, namely of the rational numbers.

10.3 Compactness

So far, our discussion of fundamentals has gone from a particular topological space, namely, real Euclidean space, to the more general topological space (which is seen in the text to be required for control applications), a 'second-countable Hausdorff' space. Requiring that kind of countability and separability, is as general as we get. This next topic of compactness refers to an additional property that is very useful for the spaces to have. Research papers often discuss a concept up to a certain point and then invoke compactness to enable satisfyingly tidy conclusions to be reached (or untidy aberrations to be avoided). Since, as a matter of fact, it is usually the tidy conclusion that raise the interest of the design engineer, it pays off in practice to look quite hard at the space one happens to be working with to see if it is compact.

In a Euclidean space, compactness is tantamount to the space being closed and bounded. A closed space contains all its limits so that sequences can end in the space. Boundedness means that every point is within reach—paths don't end at rainbows that constantly recede. The useful property abstracted from closedness and boundedness is that together they imply that not only can one count the number of open sets one needs to cover the space of concern, but also that, after a while, one finishes the counting. This is the property of compactness: only a finite number of open sets are required to cover every point of the whole space and still have it that any two points can belong to different open sets.

10.4 Continuity

It is assumed that the reader is familiar with the concept of functions defined on $\mathbf{R}^n$, and with continuity of functions as defined in terms of limits. Continuity can also be defined in a way that makes sense in arbitrary topological spaces. (From now on, we will refer to the topological spaces (X, Ω_x) and (Y, Ω_y) by just X and Y). Consider two topological spaces X and Y. Let f be a function with domain X and range Y. Let U be an open set in Y. The inverse image of U is the set of elements in X that maps into $U : f^{-1}(U) = \{x \in X : f(x) \in U\}$. Then we say that f is a *continuous function* if for each open set U in $Y, f^{-1}(U)$ is open in X. If X and Y are Euclidean spaces, then this definition in terms of open sets is equivalent to the one utilizing limits and open intervals or balls.

Continuity is a key property required throughout the text every time two spaces are related to each other. It preserves a topological consistency in relating two spaces that is expressed by the word 'homeomorphism'. Spaces are homeomorphic if they are related by a continuous function that has a continuous function as its inverse. Then the topological consistency arises from such facts as that the continuous images: of open sets are open; and of compact sets are compact. The carrying over of these structural properties is a basic requirement in differential geometry. In fact, it is generally required that various orders of derivatives of the relating functions be continuous. Then the functions are called diffeomorphisms.

10.5 Derivative

The reader will recall that the definition of the derivative of a function f with domain an open subset E of $\mathbf{R}^n$ and with range in $\mathbf{R}^m$ is slightly more complicated than in the case of a function of single variable. We say that f is differentiable at $x \in E$ if there is a linear map $A(x)$, a Jacobian matrix, from $\mathbf{R^n}$ to $\mathbf{R^m}$ such that

$$\lim_{|h|\to 0} \frac{|f(x+h) - f(x) - A(x)h|}{|h|} = 0.$$

Then $A(x)$ is called the *derivative* of f, and we will let the Jacobian matrix be called $f'(x)$; $A(x) = f'(x)$. At any particular place, x, $A(x)$ is a particular linear map from $\mathbf{R^n}$ to $\mathbf{R^m}$. Hence $A(\ )$ is a function of whose domain is E and whose range is the space of linear maps from $\mathbf{R^n}$ to $\mathbf{R^m}$ denoted by $L(\mathbf{R^n}, \mathbf{R^m})$. Now $L(\mathbf{R^n}, \mathbf{R^m})$ can be identified with the $n \times m$ matrices and hence with $\mathbf{R^{nm}}$. Thus, $A(\ )$ is a map from $\mathbf{R^n}$ to $\mathbf{R^{nm}}$ whose continuity and differentiability can be discussed.

We say that f is a $C^1(E)$ function if A, its derivative, is continuous, and we say f is a $C^2(E)$ function if A is a $C^1(E)$ function. If f is a $C^n(E)$ function for all n, we say that f is a $C^\infty(E)$ function. Note that even though this says that f has derivatives of all orders, this is a strictly weaker condition than saying that f is represented by a power series. The function

$$f(x) = \begin{cases} \exp(-1/x^2), x > 0 \\ \quad 0 \quad , \quad x \le 0 \end{cases}$$

is the classical example of a C^∞ function that is not represented by a power series at $x = 0$. A function that is represented by a convergent power series is called *analytic*. In this book, all functions, except those that are solutions of differential equations, will be assumed to be C^∞. If both f and its inverse, f^{-1}, are C^∞ functions, then f is said to be a *diffeomorphism*.

10.6 Inverse Function Theorems

The section on manifolds, the basic notion of a space in differential geometry, shows that a manifold is very closely associated with functions that are implicitly defined. The following discussion may refresh the reader's memory of such implicit functions.

Suppose one has a function f which maps $\mathbf{R}^n \times \mathbf{R}^m$ into $\mathbf{R}^m$; that is, $f : \mathbf{R}^n \times \mathbf{R}^m \to \mathbf{R}^m$;

$$
\begin{aligned}
f &= f(x_1, \cdots, x_n, y_1, \cdots, y_m) \\
&= (f_1, \cdots, f_m) \\
&= (0, \cdots, 0)
\end{aligned}
$$

The implicit function theorem gives rather general conditions under which f can be reformulated so that m of the variables, y, can be expressed explicitly as a function of the other n variables $x : y = y(x)$.

A simple example shows that the m-valued function f can be expanded to an $(n + m)$-valued function which is invertible. Let f be given by

$$
a_1 x_1 + a_2 x_2 + a_3 x_3 - b = 0
$$

Two identifies, $x_2 - x_2 = 0$ and $x_3 - x_3 = 0$ can be added trivially to form the set:

$$
\begin{bmatrix} a_1 & a_2 & a_3 \\ 0 & 1 & 0 \\ 0 & 0 & 1 \end{bmatrix}
\begin{bmatrix} x_1 \\ x_2 \\ x_3 \end{bmatrix} =
\begin{bmatrix} b \\ x_2 \\ x_3 \end{bmatrix}
$$

Then, if the determinant of the coefficient matrix is not zero, that is if $a_1 \neq 0$, then the set of equations can be solved for x_1 in terms of x_2 and x_3.

The example shows how finding the explicit function x_1 from its implicit representation $a_1 x_1 + a_2 x_2 + a_3 x_3 - b = 0$ involves finding an inverse. Hence, not only is it useful to state the Implicit Function theorem, but it also is desirable to state the Inverse Function theorem as well, to serve both as a lemma and as a special case. The two theorems will only be quoted here. The reader can consult any advanced calculus book for a discussion and interpretation of them.

Inverse and Implicit Function Theorems

Theorem 10.1 (Inverse function theorem) *In [4, p. 35], suppose that $f : \mathbf{R}^n \to \mathbf{R}^n$ is a continuously differentiable function on an open set containing the point a and suppose that the Jacobian, $\det f'(a), \neq 0$. Then there is an open set V containing a and an open set W containing $f(a)$ such that $f : V \to W$ has a continuous inverse $f^{-1} : W \to V$ which is differentiable and which for all $y \in W$ satisfies*

$$(f^{-1}(y))' = (f'(f^{-1}(y)))^{-1}$$

In the next theorem, the expression $D_j f^i(x)$ refers to the derivative along the *jth* coordinate of the *ith* component of the vector-valued function f with the value taken at the point x.

Theorem 10.2 (Implicit function theorem) *In[4, p.*

41], supposing $f : \mathbf{R}^n \times \mathbf{R}^m \to \mathbf{R}^m$ is continuously differentiable in an open set containing (a, b), and that $f(a, b) = 0$. Let M be the $m \times m$ matrix

$$D_{n+j} f^i(a, b), \ 1 \leq i, \ j \leq m$$

If $\det M \neq 0$, then there is an open set $A \subset \mathbf{R}^n$ containing a and an open set $B \subset \mathbf{R}^m$ containing b with the following property: for each $x \in A$ there is a unique $g(x) \in B$ such that $f(x, g(x)) = 0$. The function g is differentiable.

The background in advanced calculus which this appendix has recalled is a sufficient collection of information for this report. Many other small details are useful to recall, and they are pointed out as the need for them arises.

Bibliography

[1]   R. Hermann, Differential Geometry and the Calculus
      of Variations. Academic Press, New York, N.Y., 1968.

[2]   W.M. Boothby, An Introduction to Differentiable
      Manifolds and Riemannian Geometry. Academic
      Press, New York, N.Y., 1977.

[3]   L. Auslander and R.E. MacKenzie, Introduction to
      Differentiable Manifolds. Dover Publications, New
      York, N.Y., 1977.

[4]   M. Spivak, Calculus on Manifolds. W.A. Benjamin,
      Inc., New York, N.Y., 1965.

[5]   J.M.C. Clark, The Consistent Selection of Parame-
      terizations in System Identification. In Proceedings
      of the Joint Automatic Control Conference, July 27-
      30, 1976, Purdue University, West Lafayette, Indi-
      ana, pp. 576-585. Amer. Soc. of Mech. Engineers, New
      York.

[6]   R.W. Brockett, Some Geometric Questions in the
      Theory of Linear Systems. IEEE Trans. on Automatic
      Control, vol. AC-21:4, Aug. 1976, pp. 449-455.

[7] M. Hazewinkel and R.E. Kalman, On Invariance, Canonical Forms and Moduli for Linear Constant, Finite-Dimensional, Dynamical Systems. In Lecture Notes on Economics and Mathematical System Theory, vol. 131 (Springer-Verlag, Berlin, 1976), pp. 440-454.

[8] C.I. Byrnes and N.E. Hurt, On the Moduli of Linear Dynamical Systems. Adv. in Mathematics, Supplementary Series, vol. 4, 1978, pp. 83-122. Academic Press, New York.

[9] W.W. Sawyer, A First Look at Numerical Functional Analysis. In Oxford Applied Mathematics and Computing Science Series, Clarendon Press, Oxford, 1978.

[10] R. Hermann and A.J. Krener, Nonlinear Controllability and Observability, IEEE Trans. on Automatic Control, vol. AC-22:5, Oct. 1977, pp. 728-740.

[11] H. Flanders, Differential Forms with Applications to the Physical Sciences, in Mathematics in Science and Engineering, Vol. 11 (Academic Press, New York, 1963).

[12] F. Klein, Elementary Mathematics from an Advanced Standpoint, Vol. 2, Geometry. Trans. from the 3rd German ed. by Hedrick and Noble (Dover Publications).

[13] J.E. Potter, Matrix Quadratic Solutions. SIAM J. Appl. Math., vol 14:3, May 1966, pp. 496-501.

[14] R. Abraham and J.E. Marsden, Foundations of Mechanics. 2nd Edition. Benjamin/Cummings Publishers, Reading, Massachussetts, 1978.

[15]   V.I. Arnold (K. Vogtmann and A. Weinstein, transl.), Mathematical Methods of Classical Mechanics. Graduate Texts in Mathematics. Springer-Verlag, New York, N.Y., 1978.

[16]   C.I. Byrnes and C. Martin, eds., Geometrical Methods for the Theory of Linear Systems. Reidel Publishing Co., Dordrecht, Holland, 1980.

[17]   C.I. Byrnes and C. Martin, eds., Linear System Theorem. Vol. 18 of Lectures in Applied Mathematics, American Mathematical Society, Providence, R.I., 1980.

Index

159